The ESSENTIALS® of

Math for Computer Applications

Staff of Research and Education Association
Dr. M. Fogiel, Director

REA

Research & Education Association
61 Ethel Road West
Piscataway, New Jersey 08854

THE ESSENTIALS ®
OF MATH FOR COMPUTER APPLICATIONS

Printed in the United States of America

Library of Congress Catalog Card Number 98-67990

International Standard Book Number 0-87891-314-9

ESSENTIALS is a registered trademark of
Research & Education Association, Piscataway, New Jersey 08854

II-1

WHAT "THE ESSENTIALS" WILL DO FOR YOU

This book is a review and study guide. It is comprehensive and it is concise.

It helps in preparing for exams, in doing homework, and remains a handy reference source at all times.

It condenses the vast amount of detail characteristic of the subject matter and summarizes the **essentials** of the field.

It will thus save hours of study and preparation time.

The book provides quick access to the important facts, principles, theorems, concepts, and equations of the field.

Materials needed for exams, can be reviewed in summary form – eliminating the need to read and re-read many pages of textbook and class notes. The summaries will even tend to bring detail to mind that had been previously read or noted.

This "ESSENTIALS" book has been carefully prepared by educators and professionals and was subsequently reviewed by another group of editors to assure accuracy and maximum usefulness.

Dr. Max Fogiel
Program Director

CONTENTS

CHAPTER 1

LOGIC

1.1 STATEMENTS

1) A declarative sentence that is either true or false, but not both simultaneously, is called a statement. A simple statement contains one declaration, while a compound statement contains two or more declarations. We shall use lowercase letters to denote simple statements and uppercase letters to denote compound statements.

Examples: a - It is raining.
A - It is raining and it is dark.
b - Fire is cold.
B - Fire is cold and ice is warm.

2) A truth table shows relationships between statements. The letters denoting the statements are listed on top of the table, while the letters T, for "true", and F, for "false", are listed below to show the relationships.

Example: a - It is raining. b - It is not raining.

a	b
T	F
F	T

3) A connective is a structure used to connect simple statements together to form a compound statement.

a) If x is a statement, then NOT x, or the negation of x, denoted by $\sim x$, is false whenever x is true, and true whenever x is false.

Example: x - Jack is home.
~x - Jack is not home.

x	~x
T	F
F	T

b) If x and y are statements, then X AND Y, or the conjunction of x and y, denoted by x ∧ y, is true when both x and y are true. The conjunction is false if x and/or y is false.

Example: x - Jack is home.
y - Dianna is away.
x ∧ y - Jack is home and Dianna is away.

x	y	x∧y
T	T	T
T	F	F
F	T	F
F	F	F

c) If x and y are statements, then x OR y, or the disjunction of x and y, denoted by x V y, is false when both x and y are false, and true otherwise.

Example: x - Jack is home.
y - Dianna is away.
x V y - Jack is home or Dianna is away.

x	y	xVy
T	T	T
T	F	T
F	T	T
F	F	F

d) If x and y are statements, then IF x THEN y, or the conditional of x and y, denoted by x → y, is false when x is true and y is false. Otherwise, the conditional is true.

Example: x - Jack is home.
y - Dianna is away.
x → y - If jack is home, then Dianna is away.

x	y	x → y
T	T	T
T	F	F
F	T	T
F	F	T

e) If x and y are statements, then x IF AND ONLY IF y, or the biconditional of x and y, denoted by x ↔ y, is true when x and y are either both true or both false, and it is false otherwise.

Example: x - Jack is home.
y - Dianna is away.
$x \leftrightarrow y$ - Jack is home if and only if Dianna is away.

x	y	$x \leftrightarrow y$
T	T	T
T	F	F
F	T	F
F	F	T

4) Two statements x and y, simple or compound, are logically equivalent, denoted by $x \equiv y$ if and only if they have the same truth values in every logical possibility.

1.2 TAUTOLOGY, IMPLICATION, AND EQUIVALENCE

1) A tautology is a statement that is true for all logical possibilities.

Example: A - Either it is raining or it is not raining. A is a tautology.

2) If the conditional of two statements x and y, compound or simple, is a tautology, then that conditional statement is called an implication, reads as x IMPLIES y and denoted by $x \Rightarrow y$.

Example: x - It is raining
y - The streets are wet.
$x \Rightarrow y$ - If it is raining, then the streets are wet.

3) Theorem: Let x and y be any two statements. Then

a) $x \Rightarrow x \vee y$ (Law of Addition)

b) $x \wedge y \Rightarrow x$ (Laws of Simplification)

$x \wedge y \Rightarrow y$

c) $(x \vee y) \wedge (\sim x) \Rightarrow y$ (Disjunctive Syllogism)

4) If the biconditional of two statements x and y, simple or compound, is a tautology, then that statement is an equivalence. It is read x IS EQUIVALENT TO y and is denoted by $x \Leftrightarrow y$.

Example: x - Sam has no money.

y - Sam is poor.
$x \Leftrightarrow y$ - Sam has no money is equivalent to Sam is poor.

5) The terms $x \equiv y$ and $x <=> y$ have the same meaning, so the symbols $\equiv$ and $<=>$ can be used interchangeably.

6) Theorem: If x and y are any two statements, then:

a) $\sim(\sim x) \equiv x$ (Law of Double Negation)

b) $x \wedge y \equiv y \wedge x$ (Communicative Law)

$x \vee y \equiv y \vee x$

c) $x \wedge x \equiv x$ (Law of Idempotency)

$x \vee x \equiv x$

d) $x \rightarrow y \equiv \sim y \rightarrow \sim x$ (Contrapositive Law)

e) $\sim(x \wedge y) \equiv x \vee \sim y$ (DeMorgan's Laws)

$\sim(x \vee y) \equiv \sim x \wedge \sim y$

7) Theorem: If x, y and z are any three statements, then:

a) $(x \wedge y) \wedge z \equiv x \wedge (y \wedge z)$ (associative laws)

$(x \vee y) \vee z \equiv x \vee (y \vee z)$

b) $x \wedge (y \vee z) \equiv (x \wedge y) \vee (x \wedge z)$ (distributive laws)

$x \vee (y \wedge z) \equiv (x \vee y) \wedge (x \vee z)$

c) $(x \rightarrow y) \quad (y \rightarrow z) => x \rightarrow z$ (transitive laws)

8) Theorem: If w, x, y and z are any four statements, then:

a) $(w \rightarrow x) \wedge (y \rightarrow z) => (w \vee y) \rightarrow (w \vee z)$ (constructive dilemmas)

$(w \rightarrow x) \wedge (y \rightarrow z) \Rightarrow (w \wedge y) \rightarrow (w \wedge z)$

b) $(w \rightarrow x) \wedge (y \rightarrow z) \Rightarrow (\sim x \vee \sim z) \rightarrow (\sim w \vee \sim y)$

$(w \rightarrow x) \wedge (y \rightarrow z) \Rightarrow (\sim x \wedge \sim z) \rightarrow (\sim w \wedge \sim q)$

(destructive dilemmas)

9) Theorem: If x and y are any two statements, then:

a) $(x \rightarrow y) \wedge x \Rightarrow y$ (Modus Ponens)

b) $(x \rightarrow y) \wedge \sim y \Rightarrow \sim x$ (Modus Tollens)

c) $x \to y \Leftrightarrow ((x \wedge \sim y) \to (y \wedge \sim x))$ (Reductio ad Absurdum)

1.3 CONTRADICTION AND INDUCTION

1) A contradiction is a statement which is false for every logical possibility. A tautology will be denoted by t, and a contradiction will be denoted by c.
($\sim t \equiv c$, $\sim c \equiv t$)

   Example: $-P \Leftrightarrow P$ is a contradiction.

2) Theorem: If x is any statement, then:

   a) $x \wedge t \Leftrightarrow x$

      $x V t \Leftrightarrow t$

   b) $x V c \Leftrightarrow x$

      $x \wedge c \Leftrightarrow c$

   c) $c \Rightarrow x$

      $x \Rightarrow t$

3) Principle of Mathematical Induction I: Let P be a proposition defined on the positive integers N (P(n) is either true or false for every $n \in N$). Then P is true for every $n \in N$ if:

   a) P(1) is true.

   b) P(n+1) is true whenever P(n) is true.

4) Principle of Mathematical Induction II: Let P be a proposition defined on the positive integers N. Then P is true for every $n \in N$ if:

   a) P(1) is true.

   b) P(n) is true whenever P(k) is true for all $1 \leq k < N$.

1.4 VALIDITY

1) An argument is the assertion that a statement, called the conclusion, follows logically from other statements, called the premises or the hypothesis.

2) A formal proof of validity for a given argument is a sequence of statements each of which is either a premise of the argument or follows from preceding statements by a known valid argument, ending with the conclusion of the argument.

3) The indirect proof of validity for a given argument is done by including, as an additional premise, the negation of it s conclusion, and then deriving a contradiction.

CHAPTER 2

SET THEORY

2.1 SETS AND SUBSETS

1) A set is any well-defined list or collection of objects, called elements. Sets are denoted by capital letters while elements are denoted by lowercase letters.

2) The statement "is an element of" is denoted by the symbol $\in$.

3) Two sets A and B are equal if and only if they have the exact same elements.

4) A set can be specified either by listing its elements or by stating the characteristic properties of its elements.

5) Special symbols for commonly used sets:

   N is the set of positive integers $\{1,2,3,\ldots\}$
   Z is the set of all integers $\{\ldots -2,-1,0,1,2,\ldots\}$
   Q is the set of rational numbers
   R is the set of real numbers
   C is the set of complex numbers

6) The universal set, denoted by the letter U, is the set of all possible elements in the set environment being used.

7) The set with no elements is called the empty set, or null set, and is denoted by ϕ.

8) If every element in a set A is also an element of a set B, then A is called a subset of B. The relationship is written $A \subseteq B$.

9) Theorem: For any set A:

a) $\phi \subseteq A \subseteq U$

b) $A \subseteq A$

c) If $A \subseteq B$ and $B \subseteq C$, then $A \subseteq C$

d) $A=B$ if and only if $A \subseteq B$ and $B \subseteq A$

10) If $A \subseteq B$, but $A \neq B$, we say A is a proper subset of B. This relationship is written $A \subset B$.

2.2 SET OPERATIONS AND VENN DIAGRAMS

1) A Venn diagram is a pictorial representation of sets by sets of points in the plane. The universal set U is represented by the interior of a rectangle, and the other sets are represented by disks lying within the rectangle.

Fig. 2.1

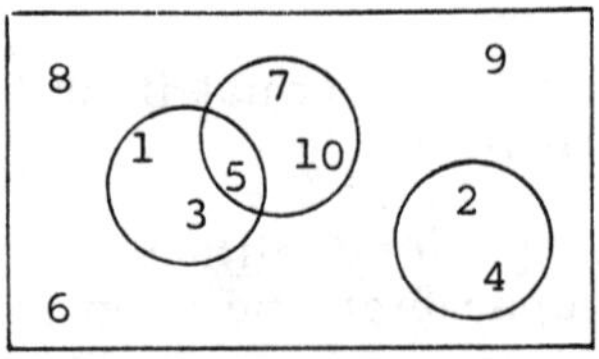

2) The union of two sets A and B, denoted by $A \cup B$, is the set of all elements which belong either to A or to B or to both A and B:

$$A \cup B = \{x \mid x \in A \text{ or } x \in B \text{ or } x \in A \text{ and } B\}$$

Fig. 2.2

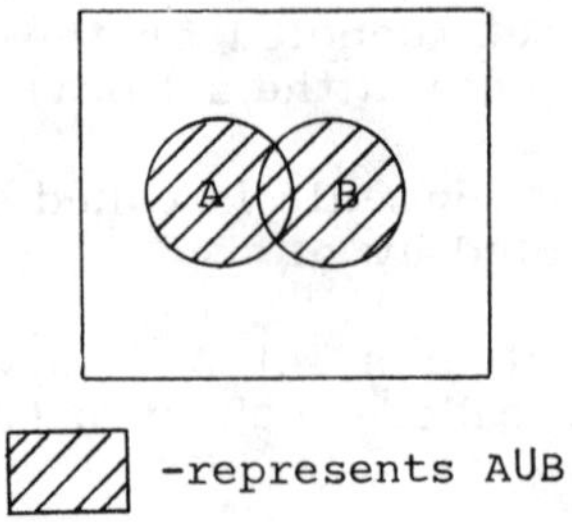

3) The intersection of two sets A and B, denoted by $A \cap B$, is the set of elements which belong to both A and B.

$$A \cap B = \{x \mid x \in A \text{ and } x \in B\}$$

Fig. 2.3

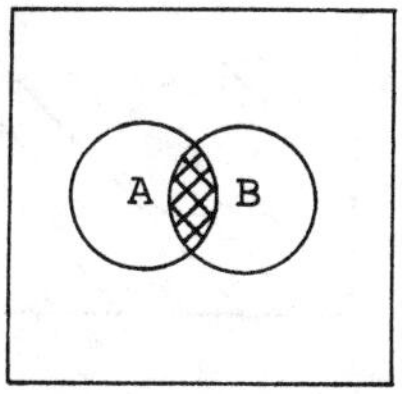

 -represents $A \cap B$

4) If $A \cap B = \phi$, that is, if A and B do not have any elements in common, then A and B are said to be disjoint.

5) The relative complement of a set B with respect to a set A, denoted by $A \setminus B$, is the set of elements which belong to A but which do not belong to B.

$$A \setminus B = \{x \mid x \in A \text{ and } x \notin B\}$$

Fig. 2.4

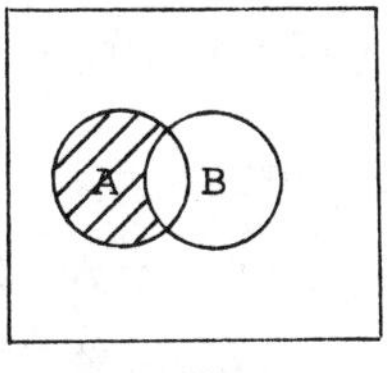

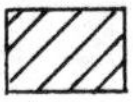 -represents A/B

6) The absolute complement of a set A, denoted by A^c, is the set of elements which belong to U but which do not belong to A.

$$A^c = \{x \mid x \in U, x \notin A\}$$

Fig. 2.5

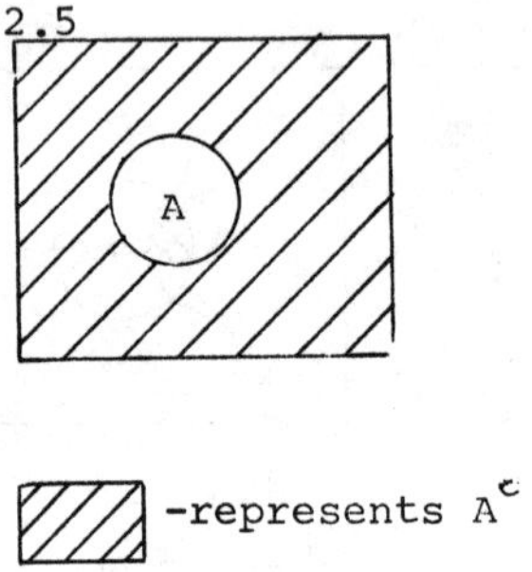

7) Theorem - The following are equivalent: $A \subseteq B$, $A \cap B = A$, $A \cup B = B$.

8) Laws of the Algebra of Sets:

<u>Idempotent Laws</u>

a) $A \cup A = A$ $\qquad$ $A \cap A = A$

<u>Associative Laws</u>

b) $(A \cup B) \cup C = A \cup (B \cup C)$ $\qquad$ $(A \cap B) \cap C = A \cap (B \cap C)$

<u>Commutative Laws</u>

c) $A \cup B = B \cup A$ $\qquad$ $A \cap B = B \cap A$

<u>Distributive Laws</u>

d) $A \cup (B \cap C) = (A \cup B) \cap (A \cup C)$ $\qquad$ $A \cap (B \cup C) = (A \cap B) \cup (A \cap C)$

<u>Identity Laws</u>

e) $A \cup \phi = A$ $\qquad$ $A \cap U = A$

$A \cup U = U$ $\qquad$ $A \cap \phi = \phi$

<u>Involution Law</u>

f) $(A^c)^c = A$

Complement Laws

g) $A \cup A^c = U$ $\qquad A \cap A^c = \phi$

$U^c = \phi$ $\qquad \phi^c = U$

DeMorgan's Laws

h) $(A \cup B)^c = A^c \cap B^c$ $\qquad (A \cap B)^c = A^c \cup B^c$

Fig. 2.6

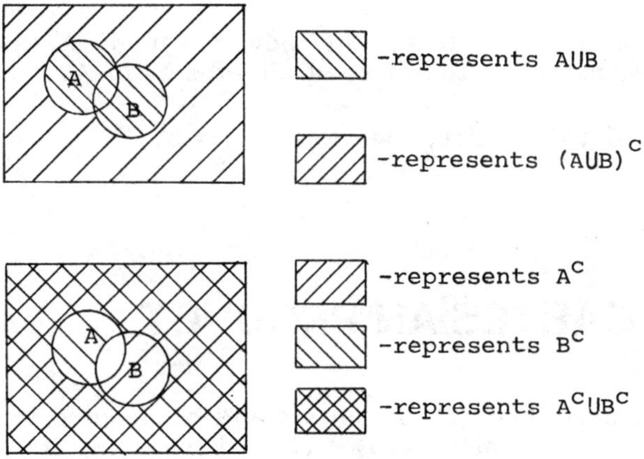

2.3 COUNTING PRINCIPLE

1) A set is said to be finite if it contains exactly m distinct elements , where m denotes some non-negative integer. Otherwise, a set is said to be infinite. If a set A is finite, we let N(A) denote the number of elements of A.

2) If A, B and C are finite sets, then:

$$N(A \cup B) = N(A) + N(B) - N(A \cap B),$$

$$N(A \cup B \cup C) = N(A) + N(B) + N(C) - N(A \cap B)$$

$$- N(A \cap C) - N(B \cap C) + N(A \cap B \cap C),$$

etc....

2.4 THE POWER SET

1) A set of sets is referred to as a class of sets. A subclass refers to some of the sets in a class.

2) For a given set A, the power set of A, denoted by P(A), refers to the class of all subsets of A.

3) If A is a finite set, then $N(P(A)) = 2^{N(A)}$.

2.5 CARTESIAN PRODUCT

1) An ordered pair of elements a and b, denoted by (a,b), is an object in which (a,b) = (b,a) if and only if a = b.

2) Let A and B be two arbitrary sets. The set of all ordered pairs (a,b), where $a \in A$ and $b \in B$, is called the cartesian product of A and B, denoted by $A \times B$.

$$A \times B = \{(a,b) / a \in A \text{ and } b \in B\}$$

3) The cartesian product can be extended over any finite number of sets:

$$A_1 \times A_2 \times \ldots \times A_N = \{(a_1, a_2, \ldots, a_N) / a_1 \in A_1, a_2 \in A_2, \ldots, a_N \in A_N\}.$$

4) The cartesian product of a set on itself is referred to in exponential notation:

$$A \times A = A^2, \quad A \times A \times A = A^3, \quad A \times A \times A \times A \times A \times A \times A = A^7$$

CHAPTER 3

RELATIONS AND FUNCTIONS

3.1 RELATIONS, INVERSE RELATIONS, COMPOSITION OF RELATIONS

1) Let A and B be sets. A relation R, from A to B, is a subset of $A \times B$. If $(x,y) \in R$, we say that x is R-related to y and denote this by xRy.

2) If R is a relation from A to A, i.e., is a subset of $A \times A$, then we say that R is a relation on A.

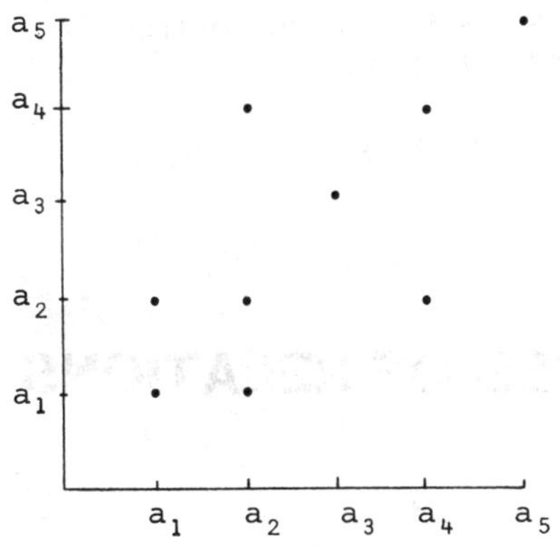

Fig. 3.1

The dots signify membership in R.
R is reflexive and symmetric, but not transitive.

3) The domain of a relation R is the set of all first elements of the ordered pairs which belong to R, and the range of R is the set of second elements.

3.1.1 INVERSE RELATIONS AND COMPOSITION OF RELATIONS

1) Let R be a relation from A to B. The inverse of R, denoted by R^{-1}, is the relation from B to A which consists of those ordered pairs which when reversed belong to R:

$$R^{-1} = \{(b,a)/(a,b) \in R\}$$

2) Theorem: If R is a relation then:

   a) $(R^{-1})^{-1} = R$

   b) domain of R = range of R^{-1}

   domain of R^{-1} = range of R

3) Let A, B and C be sets, and let R be a relation from A to B, and S be a relation from B to C. Then the composition of R and S, denoted by $R \cdot S$, is defined as:

$$R \cdot S = \{(a,c)/\text{there exists } b \in B \text{ for which } (a,b) \in R \text{ and } (b,c) \in S\}$$

4) Theorem - Let A, B, C, and D be sets. If R is a relation from A to B, S is a relation from B to C, and T is a relation from C to D, then:

$$(R \cdot S) \cdot T = R \cdot (S \cdot T)$$

3.2 PROPERTIES OF RELATIONS

1) Let R be a relation on a set A.

   a) R is reflexive if aRa for every $a \in A$.

   b) R is symmetric if aRb implies bRa.

   c) R is anti-symmetric if aRb and bRa implies a=b.

   d) R is transitive if aRb and bRc implies aRc.

   e) R is atransitive if aRb and bRc implies $a \not R c$.

2) Let S be any nonempty set. A partition of S is a collection $\{A_i\}$ of nonempty subsets of S such that:

a) Each $a \in S$ belongs to one of the A_i.

b) If $A_i \neq A_j$, then $A_i \cap A_j = \phi$.

The subsets in a partition are called cells.

3.3 EQUIVALENCE RELATIONS

1) A relation R on a set is called an equivalence relation if R is reflexive, symmetric, and transitive.

2) Let R be an equivalence relation in a set A and, for each $a \in A$, let [a], called the equivalence class of A, be the set of elements to which a is related:

$$[a] = \{X/(a,x) \in R\}$$

3) The collection of equivalence classes of A, denoted by A/R, is called the quotient of A by R:

$$A/R = \{[a]/a \in A\}$$

4) Theorem: Let R be an equivalence relation in a set A. Then:

a) $a \in [a]$ for every $a \in A$

b) $[a] = [b]$ if and only if $(a,b) \in R$.

c) if $[a] \neq [b]$, then [a] and [b] are disjoint.

3.4 FUNCTIONS

1) A function f assigns unique elements in a set B to each element in a set A, denoted by $f:A \rightarrow B$. The set A is called the domain of the function; B is called the codomain of the function.

2) For a function $f: A \to B$, read "f maps A into B", if $a \in A$, then the unique element of B which the function f assigns to a is called the image of a under f, denoted by f(a). The set of all such values is called the image of f, denoted by f(A).

3) If, for a function $f: A \to B$, $f(a)=b$, then a is called the independent variable, and b is called the dependent variable.

4) Two functions $f: A \to B$ and $g: A \to B$ are said to be equal if $f(a)=g(a)$ for every $a \in A$.

5) A function can be represented on a cartesian coordinate system by letting the x-axis represent the independent variable, and the y-axis represent the dependent variable. Such a representation is called the graph of a function.

3.5 SURJECTIVE, INJECTIVE AND BIJECTIVE FUNCTIONS

1) A function $f: A \to B$ is said to be injective, or one-to-one, if different elements in the set A have different images.

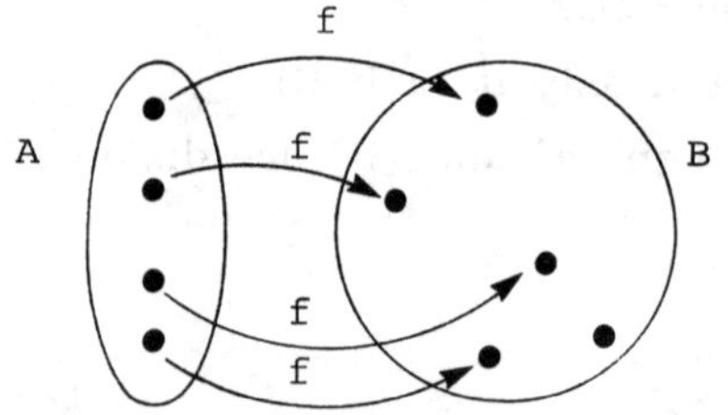

Fig. 3.2

f is 1-1, but not onto.

2) A function $f: A \to B$ is said to be surjective, or onto, if every element of B is the image of at least one element of A.

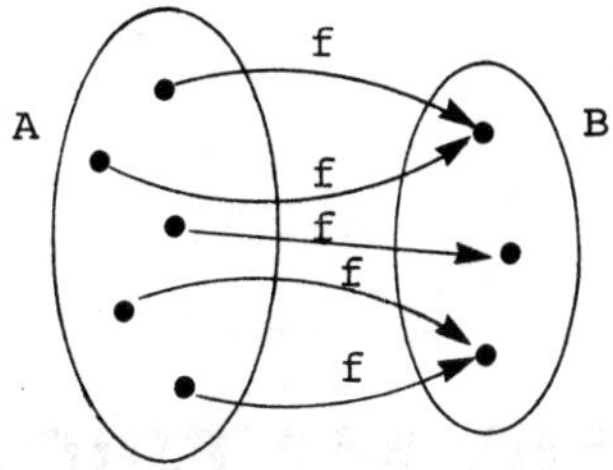

Fig. 3.3

f is onto, but not 1-1.

3) A function f:A→B is bijective if f is both surjective and injective.

4) If f is a bijective function, then so is f^{-1}.

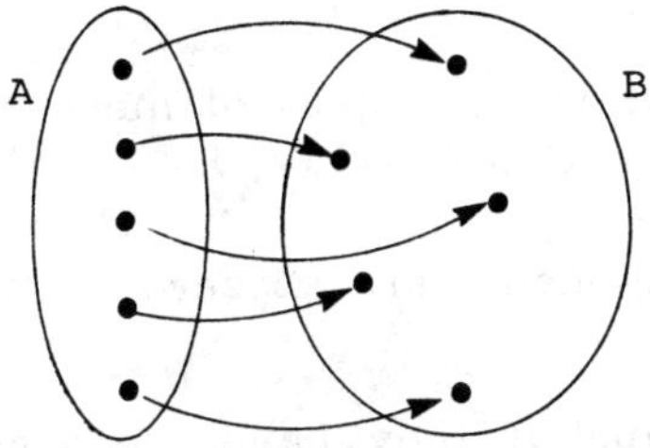

Fig. 3.4

f is a bijection.

CHAPTER 4

VECTORS AND MATRICES

4.1 VECTORS

1) A vector $\vec{u}$ is a list (or N-tuple) of numbers, called the components of $\vec{u}$. $\vec{u} = (u_1, u_2, \ldots, u_N)$

2) A vector whose components are all zero is called the zero vector ($\vec{0}$).

3) Two vectors are equal if they have the same number of components and if corresponding components are equal.

4) The sum of two vectors is the vector obtained by adding corresponding components from the two vectors.

5) The product of a scalar and a vector is the vector obtained by multiplying each component of the vector by the scalar.

6) The negative of a vector $\vec{u}$, denoted by $-\vec{u}$, is the product $-1\vec{u}$.

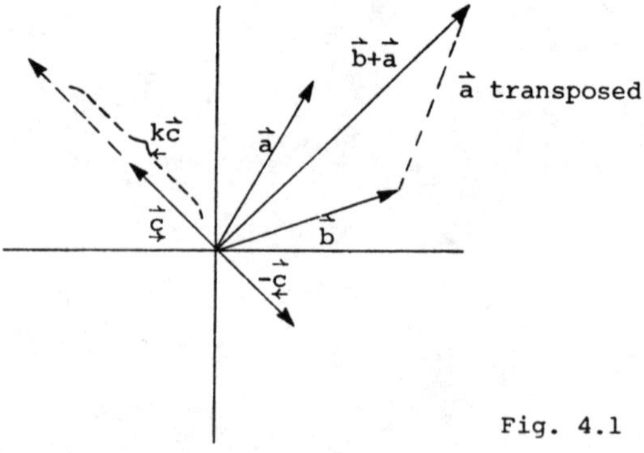

Fig. 4.1

7) The inner product of two vectors $\vec{u} = \{u_1, u_2, \ldots, u_N\}$ and $\vec{v} = \{v_1, v_2, \ldots, v_N\}$, denoted by $\vec{u} \cdot \vec{v}$, is defined by:

$$u \cdot v = u_1 v_1 + u_2 v_2 + \ldots + u_N v_N$$

8) The norm (or length) of a vector $\vec{u}$, denoted by $||\vec{u}||$, is defined by: $||\vec{u}|| = \sqrt{\vec{u} \cdot \vec{u}}$. $||\vec{u}|| = \vec{0}$ if and only if $\vec{u} = \vec{0}$.

4.2 MATRICES AND MATRIX ARITHMETIC

1) A matrix is an M×N rectangular array of numbers, called entries, where M refers to the number of rows of the matrix, and N refers to the number of columns in the matrix. Matrices are denoted by capital letters:

$$A = \begin{bmatrix} a_{11} & a_{12} & \ldots & a_{1N} \\ a_{21} & a_{22} & \ldots & a_{2N} \\ \vdots & \vdots & & \vdots \\ a_{M1} & a_{M2} & \ldots & a_{MN} \end{bmatrix}$$

2) Two matrices are equal if they have the same size, and if their corresponding entries are equal.

3) A matrix with all zero entries is called a zero matrix.

4) If A and B are two matrices of the same size, then the sum of A and B is the matrix obtained by adding their corresponding entries:

$$A+B = \begin{bmatrix} a_{11}+b_{11} & a_{12}+b_{12} & \ldots & a_{1N}+b_{1N} \\ a_{21}+b_{21} & a_{22}+b_{22} & \ldots & a_{2N}+b_{2N} \\ \vdots & \vdots & & \vdots \\ a_{M1}+b_{M1} & a_{M2}+b_{M2} & \ldots & a_{MN}+b_{MN} \end{bmatrix}$$

5) The product of a scalar k and a matrix A is the matrix obtained by multiplying each element of A by k:

$$kA = \begin{bmatrix} ka_{11} & ka_{12} & \dots & ka_{1N} \\ ka_{21} & ka_{22} & \dots & ka_{2N} \\ \vdots & \vdots & & \vdots \\ ka_{M1} & ka_{M2} & \dots & ka_{MN} \end{bmatrix}$$

6) The matrix $-1A$ is known as the negative of the matrix A.

7) (Since vectors are only $1 \times N$ matrices, the following properties apply to vectors as well as to matrices.)

Theorem: Let A, B, and C be matrices of the same size, let Z be a zero matrix, and let k and k' be scalars. Then:

a) $(A+B)+C = A+(B+C)$

b) $A+B = B+A$

c) $A+Z = A$

d) $A+(-A) = Z$

e) $k(A+B) = kA+kB$

f) $(k+k')A = kA+k'A$

g) $1A = A$

8) Let A and B be two matrices where A has the same number of columns as B has rows. That is, A is a $M \times N$ matrix, and B is a $N \times P$ matrix. Then, the product of A and B is the $M \times P$ matrix where the ijth entry of AB is obtained by multiplying the entries of the ith row of A by the corresponding elements of the jth column of B and then adding the resulting products; i.e., if AB = C, then $c_{ij} = \sum_{k=1}^{n} a_{2k} \cdot b_{kj}$.

9) Theorem: Let A, B and C be matrices which are defined for the following operations, then:

a) $(AB)C = A(BC)$

b) $A(B+C) = AB+AC$

c) $(B+C)A = BA+CA$

d) $k(AB) = (kA)B = A(kB)$, where k is a scalar.

4.3 TRANSPOSE AND INVERSE OF A MATRIX

1) The transpose of a matrix A, denoted by A^T, is the matrix obtained by making the rows of A into columns and the columns of A into rows:

If $A = \begin{bmatrix} a_{11} & a_{12} & \cdots & a_{1N} \\ a_{21} & a_{22} & \cdots & a_{2N} \\ \vdots & \vdots & & \vdots \\ a_{M1} & a_{M2} & \cdots & a_{MN} \end{bmatrix}$ then $A^T = \begin{bmatrix} a_{11} & a_{12} & \cdots & a_{M1} \\ a_{12} & a_{22} & \cdots & a_{M2} \\ \vdots & \vdots & & \vdots \\ a_{1N} & a_{2N} & \cdots & a_{MN} \end{bmatrix}$

2) Theorem: Properties of matrix transpose.

a) $(A+B)^T = A^T + B^T$

b) $(kA)^T = kA^T$ where k is a scalar

c) $(AB)^T = B^T A^T$

d) $(A^T)^T = A$

4.3.1 INVERSE OF A MATRIX

1) A matrix with the same number of rows as columns is called a square matrix.

2) In a square matrix,

main diagonal

$$A = \begin{bmatrix} a_{11} & a_{12} & \cdots & a_{1M} \\ a_{21} & a_{22} & \cdots & a_{2M} \\ \vdots & \vdots & & \vdots \\ a_{M1} & a_{M2} & \cdots & a_{MM} \end{bmatrix}$$

The main diagonal consists of the elements $(a_{11}, a_{22}, \ldots, a_{MM})$.

3) The N×N square matrix with 1's along the main diagonal and zeros everywhere else is called the unit matrix. This is denoted by I.

4) Theorem: For any square matrix A:

a) AI = A

b) $A^0 = I$

5) A square matrix A is said to be invertible if there exists a matrix B with the property that

$$AB = BA = I.$$

Matrix B is called the inverse of A and is denoted by A^{-1}.

4.4 DETERMINANTS

1) Each square matrix A has a specific number associated with it, called the determinant of A, denoted by det(A).

2) For a 1×1 matrix, the determinant is the one entry:

$$\det[a_{11}] = a_{11}.$$

For a 2×2 matrix, the determinant is defined as:

$$\det\begin{bmatrix} a_{11} & a_{12} \\ a_{21} & a_{22} \end{bmatrix} = (a_{11}a_{22}) - (a_{12}a_{21}),$$

(Refer to the Linear Algebra study guide for information on how to find the determinant of a matrix of size greater than 2×2).

3) Theorem: For any two square matrices A and B,

$$\det(AB) = \det(A) \cdot \det(B)$$

4) Theorem: A matrix is invertible if and only if it has a non-zero determinant.

CHAPTER 5

GRAPH THEORY

5.1 GRAPHS AND MULTIGRAPHS

1) A graph G consists of two things:
   a) A set V whose elements are called vertices, points, or nodes.
   b) A set E of unordered pairs of distinct vertices called edges.

   A graph is denoted by G(V,E).

2) Two vertices U and V in a graph are said to be adjacent if there is an edge {U,V} in the graph.

3) Graphs are represented by diagrams in which each node is represented by a dot, and each edge is represented by a line connecting two nodes which are called the endpoints of the edge.

4) A graph containing multiple edges (two or more edges connecting the same endpoints) and/or loops (an edge whose endpoints are the same vertex) is called a multigraph.

Fig. 5.1

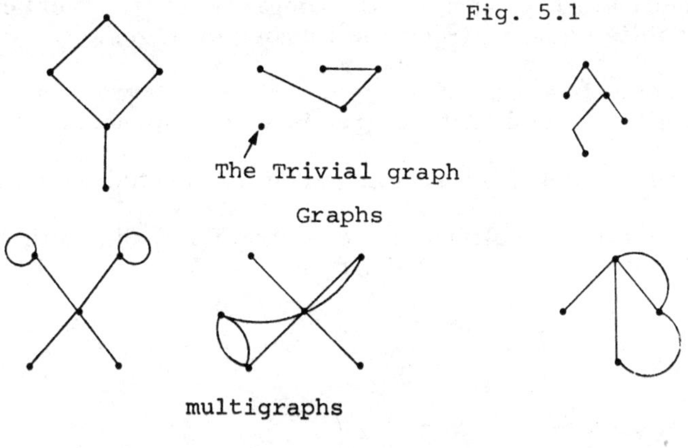

5) If G(V,E) is a graph and G(V',E') is a graph where V' is a subset of V and E' is a subset of E, then G(V',E') is called a subgraph of G(V,E). If E' contains all the edges of E whose endpoints lie in V', then G(V',E') is called the subgraph generated by V'.

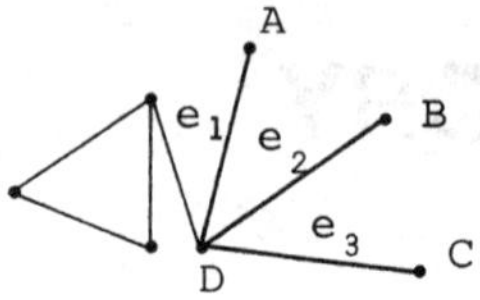

V' = {A,B,C} Fig. 5.2
E' = $\{e_1, e_2, e_3\}$
G(V',E') is the subgraph generated by V'.

6) A graph is said to be finite if it has a finite number of nodes and edges. The graph consisting of 1 vertex and no edges is called the trivial graph.

5.2 CONNECTIVITY

1) If v is an endpoint of an edge e, then e is said to be incident on v.

2) The degree of a vertex v is defined to be the number of edges which are incident on v.

3) Theorem: The sum of the degrees of the vertices of a graph is equal to twice the number of edges.

4) A node is said to be even if its degree is an even number, or odd if its degree is an odd number.

5) A vertex of degree zero is called an isolated vertex.

6) A walk in a multigraph consists of an alternating series of nodes and edges of the form:

$$v_0, e_1, v_1, e_2, v_2, \ldots, e_{N-1}, v_{N-1}, e_N, v_N$$

where each e_i is incident on v_{i-1} and v_i. The number N is called the length of the walk.

7) A walk is said to be closed if its starting point and endpoint are the same. Otherwise, the walk is said to connect the starting point to the endpoint.

8) a) A walk in which all edges are distinct is called a trail.

   b) A path is a walk in which all nodes are distinct (all paths are trails).

   c) A cycle is a closed walk in which all vertices are distinct except for the starting and ending nodes. A cycle of length k is called a k-cycle. A cycle must have a length of three or more.

9) Theorem: There is a walk from a node u to a node v if and only if there is a path from u to v.

10) A graph is said to be connected if there is a path between any two of its vertices.

Fig. 5.3

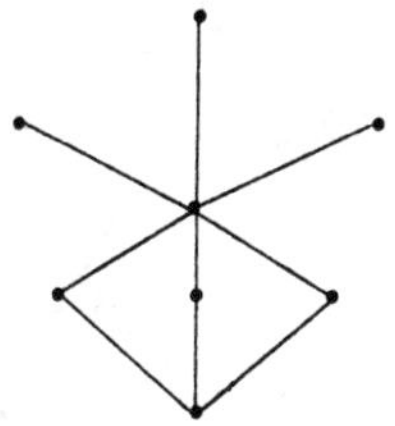

Connected graph

11) If G(V,E) is a graph, and G(V',E') is a connected subgraph of G(V,E), then G(V',E') is called a connected component of G(V,E). This is true only if G(V,E) is not contained in any larger connected subgraph.

12) The distance between two vertices u and v of a connected graph is the length of the shortest path between these two vertices.

13) The diameter of a connected graph is the maximum distance between any two of its vertices.

14) If V is a node of the graph G, then G-V is the graph

obtained by deleting V and all edges incident on V. V is called a cut point if G-V is disconnected.

5.3 TRAVERSABLE MULTIGRAPHS

1) A mutigraph is said to be traversable if there exists a walk which includes all of the nodes and uses each edge exactly once. If such a walk exists, it is called a traversable trail.

2) A graph G is called a eulerian graph if there exists a closed traversable trail, called a eulerian trail.

Fig. 5.4

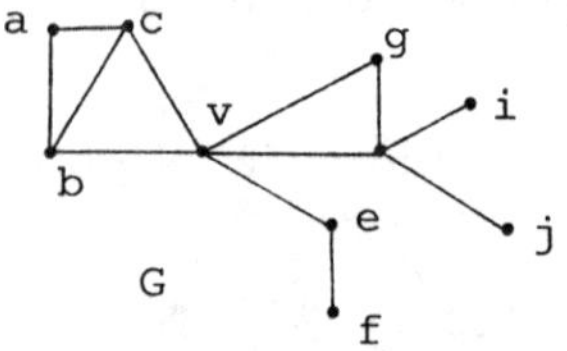

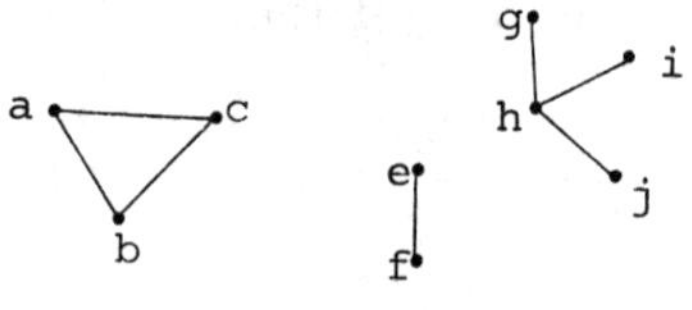

G-V

Fig. 5.5

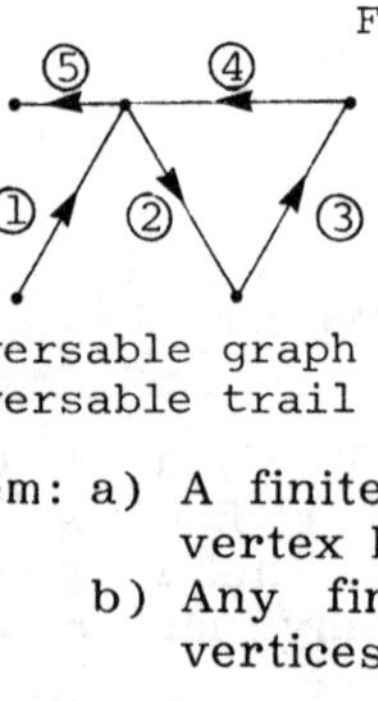

A traversable graph with a traversable trail

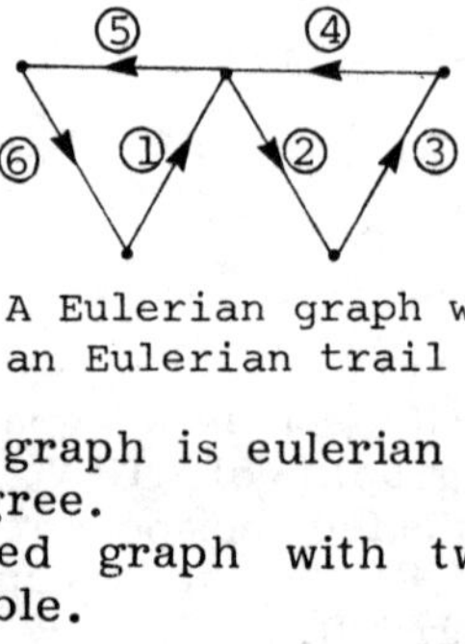

A Eulerian graph with an Eulerian trail

3) Theorem: a) A finite connected graph is eulerian if each vertex has even degree.
   b) Any finite connected graph with two odd vertices is traversable.

4) A hamiltonian cycle is a closed walk which includes each mode exactly once. A graph containing such a walk is called a hamiltonian graph.

Fig. 5.6

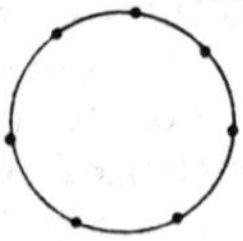

A Hamiltonian graph

5.4 GRAPH TYPES

1) A graph is complete if each node is connected to every other node.

2) A graph is k-regular if every node has degree k.

Fig. 5.7

A complete graph which is 4-regular

3) A graph G is bipartite if its nodes V can be divided into two subsets M and N such that each edge of G connects a node of M to a node of N.

Fig. 5.8

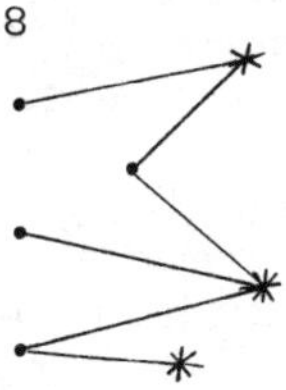

A bipartite graph nodes of M denoted by • ,those of N by *

4) A graph is acyclic if it has no cycle.

5.5 GRAPHS AND MATRICES

1) In an edge matrix B, corresponding to a graph G, each row of B consists of the starting node and the end node of an edge of 6. An edge matrix, therefore, is an N×2, where N is the number of edges in G.

2) The adjacency matrix A, with elements a_{ij}, corresponding to a graph G, is defined by an m×m matrix whose entries are defined as the following:

a_{ij} = the number of different edges e with endpoints x_i and x_j.

3) The incidence matrix M, with elements M_{ij}, corresponding to a graph G, is defined by

$$M_{ij} = \begin{cases} 1 \text{ if the node } v_i \text{ is incident on the edge } e_j \\ 0 \text{ otherwise} \end{cases}$$

4) A labeled graph is one in which the vertices and/or edges are assigned a data value. If that data value is a non-negative number assigned to an edge of the graph, then that number is called the weight, or length, of that edge.

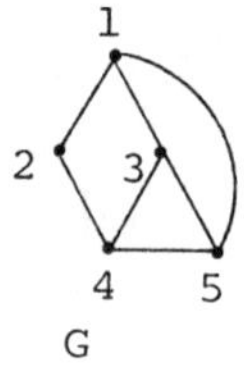

Fig. 5.9

5) Two graphs are isomorphic if there exists a one-to-one correspondence, ϕ, between the vertices of these two graphs such that ϕ preserves adjacency, i.e. if u and w are adjacent, then $\phi(u)$ and $\phi(w)$ are adjacent.

6) An elementary subdivision of a graph G is obtained by taking an edge and adding an intermediate vertex so that the two old vertices are connected via the new one.

7) One obtains a subdivision of a graph by taking successive elementary subdivisions.

8) A graph G_1 is homeomorphic from another graph G_2 if G_1 is isomorphic to G_2 or to a subdivision of G_2.

9) Two graphs are homeomorphic if there exists a third graph such that the two are both homeomorphic from the third.

5.6 PLANAR GRAPHS, COLORATIONS AND TREES

1) A graph or multigraph which can be drawn in the plane so that its edges do not cross is said to be planar.

Fig. 5.10

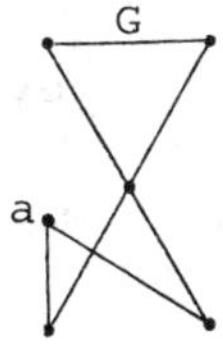

=

G is a planar graph
since it can be redrawn.

2) A planar representation of a finite planar multigraph is called a map. The map divides the plane into regions.

3) Theorem: The sum of the degrees of the regions of a map is equal to twice the number of edges.

4) Theorem (Euler's Formula): If V is the number of vertices, E is the number of edges, and R is the number of regions in any connected map, then $V-E+R = 2$.

5) Theorem: Let G be a connected planar graph with p vertices and q edges, where $p \geq 3$. Then $q \leq 3p-6$.

6) A coloring of a graph G is an assignment of colors to the vertices of G such that adjacent vertices have different colors. G is N-colorable if there exists a coloring of G which uses N colors.

7) The minimum number of colors needed to color a graph G is called the chromatic number of G and is denoted by $\chi(G)$

8) Theorem: The following are equivalent for a graph G:

   a) G is 2-colorable.

   b) G is bipartite.

   c) Every cycle of G has even length.

9) A tree is a connected graph with no cycles. A forest is a graph with no cycles.

10) The tree consisting of a single vertex with no edges is called the degenerate tree.

11) Theorem: Let G be a graph with more than one vertex. Then the following are equivalent:

a) G is a tree.

b) Each pair of vertices is connected by exactly one path.

c) G is connected, but if any edge is deleted, then the resulting graph is not connected.

d) G is cycle-free, but if any edge is added to the graph, then the resulting graph has exactly one cycle.

12) Let G be a finite graph with $N > 1$ vertices. Then the following are equivalent:

a) G is a tree.

b) G is cycle-free and has $N-1$ edges.

c) G is connected and has $N-1$ edges.

13) Trees and forests are 2-colorable.

14) A subgraph S of a graph G is called a spanning tree of G if S is a tree and includes all the vertices of G.

CHAPTER 6

COUNTING AND BINOMIAL THEOREM

6.1 PERMUTATIONS

1) A permutation of the elements of a set is a unique arrangement of those elements with no omissions or repetitions.

2) If S is a set and $N(S) = k$, then the number of permutations is $(k)(k-1)(k-2)\ldots(1)$.

3) If S is a set such that $N(S) = k$, then the number of ordered subsets of the elements of S, each containing p elements, is $(k)(k-1)(k-2)\ldots(k-p+1)$.

6.2 FACTORIALS

1) For a non-negative integer N, the factorial of N, denoted by N!, is defined by:

   a) for $N=0$, $0! = 1$

   b) for $N \neq 0$, $N! = N(N-1)(N-2)(N-3)\ldots 3 \cdot 2 \cdot 1$

2) Theorem: $N! = N \cdot (N-1)!$

3) Theorem: $N(N-1)(N-2)\ldots(N-k+1) = \dfrac{N!}{(N-k)!}$

6.3 COMBINATIONS AND THE BINOMIAL THEOREM

1) The number of combinations of r elements taken from a set of N elements, denoted by $\binom{N}{r}$, is defined by

$$\binom{N}{r} = \frac{N!}{r!(N-r)!} .$$

2) Theorem: If a set S contains N elements, k_1 of one kind, k_2 of another kind, ..., k_N of another kind, then the number of distinct arrangements that can be made of the N elements is

$$\frac{N!}{(k_1!)(k_2!)\ldots(k_N!)}$$

6.3.1 THE BINOMIAL THEOREM

1) Any expression of the form $(x+y)^N$, where x and y represent terms, and N is a non-negative integer, is called a binomial.

2) Any binomial can be expanded as follows:

$$(x+y)^N = \binom{N}{0}x^N + \binom{N}{1}x^{N-1}y + \binom{N}{2}x^{N-2}y^2 + \ldots + \binom{N}{N}y^N$$

The coefficients of $(a+b)^n$ can be expressed as Pascal's Triangle as shown below.

$(a+b)^0$	1
$(a+b)^1$	1 1
$(a+b)^2$	1 2 1
$(a+b)^3$	1 3 3 1
$(a+b)^4$	1 4 6 4 1
$(a+b)^5$	1 5 10 10 5 1
$(a+b)^6$	1 6 15 20 15 6 1
$(a+b)^7$	1 7 21 35 35 21 7 1
$(a+b)^8$	1 8 28 56 70 56 28 8 1
$(a+b)^9$	1 9 36 84 126 126 84 36 9 1

CHAPTER 7

PROBABILITY

7.1 THE PROBABILITY MEASURE

1) The probability measure of an event is a quantitative prediction of the chance of the event occuring. An event is represented by a statement, which is assigned a probability.

2) We assume that any two equivalent statements will be assigned the same probability.

3) Three steps to assigning a probability to a statement S:

   a) Determine the set of all the logical possibilities concerning the event in question, denoted by U.

   b) Assign a probability to each subset of U $(u_1, u_2, \ldots, u_N)$, denoted by $P(u_i)$, where $0 \le P(u_i) \le 1$ and where $P(u_1) + P(u_2) + \ldots + P(u_N) = 1$.

   c) Assign a probability to the statement, denoted by P(S), by adding the probabilities of all of the subsets of U which imply S.

4) Theorem: If x and y are any two statements, then:

   a) $P(x)=0$ if and only if x is false.

   b) $0 \le P(x) \le 1$

   c) $P(x \cup y) = P(x) + P(y) - P(x \cap y)$

   d) $P(\sim x) = 1 - P(x)$ or $P(\bar{x}) = 1 - P(x)$

   ($\bar{x}$ - not x, same as $\sim x$)

5) The odds on an event is a ratio of r to s, denoted by r:s, where r is a prediction of the number of times the event will occur in r+s chances.

6) The equiprobable probability measure is a probability calculated on the belief that all of the logical possibilities of the event in question have equal probabilities.

7.2 CONDITIONAL PROBABILITY

1) The conditional probability of a statement p given q, denoted by P[p/q], is the probability of p given that statement q is true.

2) $$P[p/q] = \frac{P(p \cap q)}{P(q)}$$

3) Two statements r and q are independent if:

$$P(r \cap q) = P(r) \cdot P(q)$$

7.3 STOCHASTIC PROCESSES

1) A stochastic process is a probabilistic analysis of a sequence of experiments. If the set of possible outcomes for each experiment is finite, we say that the process is finite.

2) Three types of stochastic processes:

   a) For an independent process, the probability of a certain outcome of the Nth experiment is independent of the first N-1 experiments.

b) An independent trials process is an independent process in which the probability of a certain outcome is the same in each successive experiment.

c) A Markov chain process is one in which the outcome of an experiment depends only on the outcome of the immediately preceding experiment.

3) Theorem: Let N be the number of experiments in an independent trials process with possible outcomes a,b,...,h. Let g(a) be the number of times a occurs, g(b) be the number of times b occurs, etc. Let $r_1, r_2, \ldots, r_h$ be a set of non-negative integers such that $r_1 + r_2 + \ldots + r_h = N$. Then:

$$P[(g(a)=r_1) \cap (g(b)=r_2) \cap \ldots \cap (g(h)=r_h)]$$

$$= \binom{N}{r_1, r_2, \ldots, r_h} (p(a))^{r_1} \cdot (p(b))^{r_2} \cdot \ldots \cdot (p(h))^{r_h}.$$

4) Theorem: If an experiment with two outcomes, a and b, occurs with probabilities p and 1-p, respectively, and is repeated N independent times, with g(a) being the number of times a occurs, then:

$$\boxed{P(g(a)=r) = \binom{N}{r} p^r (1-p)^{N-r}}$$

7.4 MEAN AND VARIANCE OF A FUNCTION

1) Let f be a function defined on a numerical possibility set, so that each f(x) represents a possible numerical outcome with P(x) being its associated probability. Then the mean value of f, denoted by M[f], is defined as:

$$\boxed{M[f] = \sum_x f(x) \cdot P(x)}$$

2) Theorem: Let f and g be any two functions defined on the same numerical possibility set. If k is a constant, then:

a) $M[f+g] = M[f] + M[g]$

b) $M[kf] = kM[f]$

c) $M[k] = k$

3) Theorem: Let f and g be any two probabilistically independent functions. Then $M[f \cdot g] = M[f] \cdot M[g]$.

7.4.1 VARIANCE OF A FUNCTION

1) Let f be a function defined on a numerical possibility set. Then the variance of f, denoted by V[f], is defined as:

$$V[f] = M[(f-M[f])^2]$$

2) Theorem: If f is a function defined on a numerical possibility set, and c is any constant, then:

a) $V[cf] = c^2 V[f]$

b) $V[f+c] = V[f]$

c) $V[c] = 0$

d) $V[f] = M[f^2] - (M[f])^2$

3) Theorem: If $f_1, f_2, \ldots, f_N$ is a set of probabilistically independent functions, then:

$$V[f_1 + f_2 + \ldots + f_N] = V[f_1] + V[f_2] + \ldots + V[f_N].$$

4) Theorem: Let $f_1, f_2, \ldots, f_N$ be a set of N probabilistically independent functions. Let $a = M[f_i]$ and $b = \sqrt{V[f_i]}$ for all i. Let $s_N = f_1 + f_2 + \ldots + f_N$, and $h_N = \frac{s_N}{N}$ Then:

a) $M[s_N] = Na$

b) $V[s_N] = Nb^2$

c) $M[h_N] = a$

d) $V[h_N] = b^2/N$

5) Theorem: Let "a" be a possible outcome for an experiment, and $P(a) = p$. The experiment is repeated independently N times. Let s_N be the number of times a occurs, and $h_N = \frac{s_N}{N}$ Then:

a) $M[s_N] = Np$

b) $V[s_N] = Np(1-p)$

c) $M[h_N] = p$

d) $V[h_N] = \frac{p(1-p)}{N}$.

6) If f is a function defined on a possibility set, then the standard deviation of f, denoted by D[f], is defined to be:

$$D[f] = \sqrt{V[f]}$$

7) Let f be a function with a mean of μ and a standard deviation of σ. Then the standardized function of f, denoted by f*, is defined as:

$$f^* = \frac{f-\mu}{\sigma}$$

8) For any standardized function f*:

a) $M[f^*] = 0$

b) $V[f^*] = 1$

7.4.2 THE LAW OF LARGE NUMBERS

1) Chebyshev's inequality: Let f be a function defined on a possibility set. For any positive number k:

$$P(|f-M[f]| \geq k) \leq \frac{V[f]}{k^2}$$

2) The law of large numbers - Let $f_1, f_2, \ldots, f_N$ be probabilistically independent functions with $M[f_i] = a$ and $V[f_i] = b$ for all i. Let $h_N = (f_1+f_2+\ldots+f_N)/N$, and k be a scalar. Then:

a) $P(|h_N - a| \geq k) \leq b/Nk^2$

b) $P(|h_N - a| > k) \to 0$. (as N approaches infinity)

7.5 THE POISSON APPROXIMATION

1) Consider an independent trials process with two possible outcomes, 0 or 1. Let j be the number of times a '1' occurs in N experiments. If p is the probability of an occurence of a '1' in any one experiment, and q = 1-p, then the binomial probability, denoted by w(j;N,p), is defined as:

$$w(j;N,p) = \binom{N}{j} p^j q^{N-j}$$

2) Theorem: If, for the binomial probability w(j;N,p), N approaches infinity and p approaches 0, in such a way that Np remains constant, then:

$$\lim_{N\to\infty} w(j;N,p) = \frac{(N \cdot p)^j e^{-M}}{j!}$$

3) Consider a series of experiments where the outcome in each experiment can be any non-negative integer. Let M be the given mean of the outcome set. The Poisson approximation, denoted by w(j;m), gives a probability measure for any integer occuring j times in the entire series of experiments. It is defined as:

$$w(j;m) = \frac{m^j e^{-M}}{j!}$$

4) Central Limit Theorem: Let $f_1, f_2, \ldots, f_N$ be the numerical outcome functions for an independent trials process with

$V[f_1] > 0$. If $sN^* = \dfrac{(f_1+f_2+\ldots+f_N) - M[f_1]\cdot N}{\sqrt{V[f_1]}\cdot\sqrt{N}}$, and $c \leq d$ are any two numbers, then:

$$P(c \leq s_N^* \leq d) \to \frac{1}{\sqrt{2\pi}} \int_c^d e^{-x^2/2}\,dx$$

as N approaches infinity.

5) The graph of the function $\frac{1}{\sqrt{2\pi}}\, e^{-x^2/2}$ is called the normal curve. $P(c \leq s_N^* \leq d)$ will be the area under this curve between c and d.

CHAPTER 8

BOOLEAN ALGEBRA

8.1 BOOLEAN ALGEBRA

1) Boolean Algebra - Algebra of logic.

2) The operations performed in Boolean Algebra are

   i) AND denoted by "•"

   ii) OR denoted by "+"

   iii) NOT denoted by "-" above the letter symbol.

   OR means inclusive OR; that is, X or Y means X or Y or both. Exclusive OR, denoted by +, means X or Y, but not both.

3) A Boolean algebra variable can have only one of two values, 0 or 1.

   0 is used to denote False.

   1 is used to denote True.

4) Consider the compound statement:

   The applicant is a male AND the applicant is married.

   Let X = the applicant is a male

   Y = the applicant is married

   The statement can now be written X AND Y. When is X AND Y true, and when is X AND Y false?

   X AND Y is true only if X is true and Y is true.

X	Y	X AND Y
True	True	True
True	False	False
False	True	False
False	False	False

X	Y	X • Y
1	1	1
1	0	0
0	1	0
0	0	0

*1 means True
0 means False

5) The applicant is a male OR the applicant is married.

X = the applicant is a male

Y = the applicant is married

The statement can now be written X OR Y. When is X OR Y true, and when is X OR Y false?

X OR Y is true when X is true or when Y is true or when both are true.

X	Y	X OR Y
True	True	True
True	False	True
False	True	True
False	False	False

X	Y	X + Y
1	1	1
1	0	1
0	1	1
0	0	0

6) NOT

The applicant is NOT male.

X = The applicant is male.

X	NOT X
True	False
False	True

X	$\overline{X}$
1	0
0	1

7) Summary of Relationships explained so far:

X = 1 or else	X = 0
$1 \cdot 1 = 1$ $1 \cdot 0 = 0 \cdot 1 = 0$ $0 \cdot 0 = 0$	$0 + 0 = 0$ $0 + 1 = 1 + 0 = 1$ $1 + 1 = 1$
$\overline{1} = 0$	$\overline{0} = 1$

8.1.1 SOME DEFINITIONS

1) Variables: The different letters used in a Boolean expression.

   Example: $A \cdot \overline{B} + \overline{A} \cdot C + A \cdot (D+E)$

   In this example, the number of variables is 5, (A,B,C,D,E).

2) Literal: Each occurrence of a variable or its complement.

   Example: $A \cdot \overline{B} + \overline{A} \cdot C + A \cdot (D+E)$

   would normally be written

   $A\overline{B} + \overline{A}C + A(D+E)$.

   The "•" is used only where additional clarity is required. The number of literals is 7.

3) Complement: The complement of a Boolean expression is obtained by:

   changing all •'s to +'s,
   changing all +'s to •'s,
   changing all 1's to 0's,
   changing all 0's to 1's,
   and complementing each literal.

   Example: The complement of

   $1 \cdot A + \overline{B}C + 0$ is $(0+\overline{A}) \cdot (B+\overline{C}) \cdot 1$

4) Dual: The dual of a Boolean expression is obtained by:

   changing all •'s to +'s,
   changing all +'s to •'s,
   changing all 1's to 0's,
   changing all 0's to 1's,
   but not complementing any literal.

Example: The dual of $1 \cdot A + \overline{B}C + 0$ is

$$(0+A) \cdot (\overline{B}+C) \cdot 1$$

5) The following is a list of the basic theorems of Boolean algebra, with their duals.

1a. $A+A = A$ b. $A \cdot A = A$	Idempotent Laws
2a. $A+B = B+A$ b. $A \cdot B = B \cdot A$	Commutative Laws
3a. $A+(B+C) = (A+B)+C$ b. $A \cdot (B \cdot C) = (A \cdot B) \cdot C$	Associative Laws
4a. $A \cdot (B+C) = (A \cdot B)+(A \cdot C)$ b. $A+(B \cdot C) = (A+B) \cdot (A+C)$	Distributive Laws
5a. $A+1 = 1$ b. $A \cdot 0 = 0$	Law of Union Law of Intersection
6a. $A \cdot 1 = 1$ b. $A+0 = A$	
7. $\overline{\overline{A}} = A$	Double Negative or Involution Law
8a. $A+\overline{A} = 1$ b. $A \cdot \overline{A} = 0$	Law of Complement
9a. $\overline{A+B} = \overline{A} \cdot \overline{B}$ b. $\overline{A \cdot B} = \overline{A}+\overline{B}$	DeMorgan's Laws
10. $\overline{1} = 0 \quad \overline{0} = 1$	
11a. $AB + A\overline{B} = A$ b. $(A+B)(A+\overline{B}) = A$	
12a. $A + AB = A$ b. $A \cdot (A+B) = A$	
13a. $AB + A\overline{B}C = AB + AC$ b. $(A+B) \cdot (A+\overline{B}+C) = (A+B) \cdot (A+C)$	
14a. $AB + \overline{A}C + BC = AB + \overline{A}C$ b. $(A+B) \cdot (\overline{A}+C) \cdot (B+C) = (A+B) \cdot (\overline{A}+C)$	

8.1.2 SPECIAL FORMS OF BOOLEAN EXPRESSIONS:

Forms of Particular Interest

1) Expanded sum of products
2) Expanded product of sums

3) "Minimum" sum of products

4) "Minimum" product of sums

1) Expanded Sum of Products:

To obtain the expanded sum of products from a sum of products, put in the missing variables in all possible combinations to each product.

Example: Given $\bar{A}C\bar{D} + A\bar{B}D + A\bar{C}$

Expanded sum of products is

$\bar{A}\bar{B}\bar{C}\bar{D}+\bar{A}BC\bar{D}+A\bar{B}\bar{C}D+A\bar{B}CD+A\bar{B}\bar{C}\bar{D}+AB\bar{C}\bar{D}+AB\bar{C}D$

2) Expanded Product of Sums:

To obtain the expanded product of sums from a product of sums, put in the missing variables in all possible combinations to each sum.

Given $(\bar{A}+\bar{B}+\bar{C})(\bar{A}+\bar{C}+D)(A+C)(A+\bar{D})$

Expanded product of sums is

$(\bar{A}+\bar{B}+\bar{C}+\bar{D})(\bar{A}+\bar{B}+\bar{C}+D)(\bar{A}+B+\bar{C}+D)(A+\bar{B}+C+\bar{D})(A+\bar{B}+C+D)$

$(A+B+C+\bar{D})(A+B+C+D)(A+\bar{B}+\bar{C}+\bar{D})(A+B+\bar{C}+\bar{D})$.

3) Minimum Sum of Products:

A minimum sum of products can be obtained from a sum of products by the application of the basic theorems.

4) Minimum Product of Sums:

Method for obtaining product of sums:

1) Obtain a sum of products form by multiplying out (if necessary).

2) Obtain the dual expression, which will be in a product of sums form.

3) Multiply out the dual to get it in a sum of products form.

4) Finally, obtain the dual of the dual to get a product of sums equivalent to the original.

8.1.3 FURTHER MINIMIZATION TECHNIQUES: (KARNAUGH MAP)

A Karnaugh Map: Pictorial device for the simplification and minimization of Boolean algebraic expressions involving at most six variables.

A map for n variables contains 2^n squares, there being a square on the map for every possible input combination.

Example: 2 variable map

B \ A	0	1
0	1	
1		

The '1' in the first box represents the term $\bar{A}\,\bar{B}$

Example: 3 variable map

C \ AB	00	01	11	10
0				
1				

The following is an illustration of the use of a Karnaugh map. Given the expression

$$E = XY + \bar{X}Y + \bar{X}\,\bar{Y}$$

1) Represent each term in the expression by a check in the appropriate box as shown below:

y \ x	0	1
0	✓	
1	✓	✓

2) Circle the adjacent checks in powers of two until all the eligible checks have been circled.

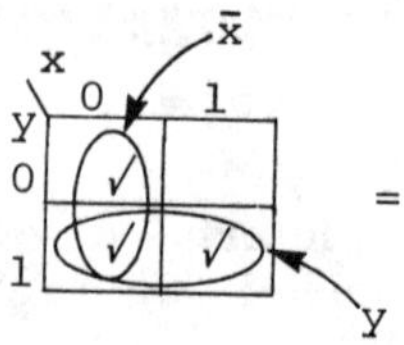

$= \bar{x} + y$

In this example, there are two groups of two 1-squares each, one group defined by $\bar{X}$ and the other by Y.

Note that every 1-square must be accounted for at least once. A 1-square may be used in as many groups as desired. Furthermore, the groups should be made as large as possible. That is, a 1-square should not be accounted for by itself if it can be accounted for in a group of two 1-squares.

Switching Circuits: Consists of switches that can either be on or off and wires which connect the switches.

A switch is denoted by the following symbol ___/ ____.

open or off   closed or on

Switches can be connected in series or parallel or complex combinations of series and parallel in such a manner as to produce the required logical function.

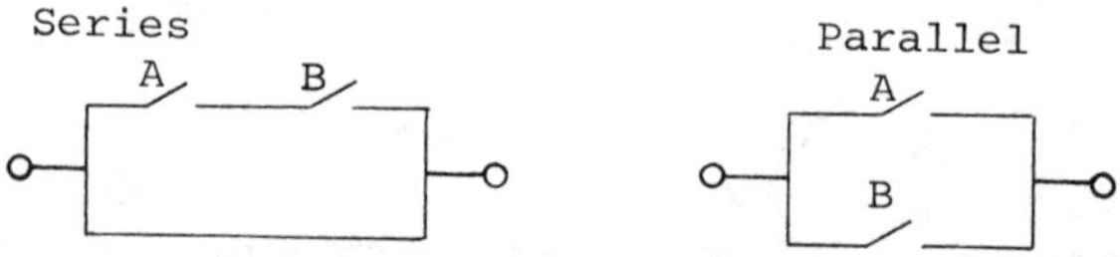

Wiring two switches in series is equivalent to performing the AND operation, or finding the logical sum of the two switches.

Wiring two switches in parallel is equivalent to performing the OR operation, or finding the logical product of the two switches.

CHAPTER 9

LINEAR PROGRAMMING AND GAME THEORY

9.1 LINEAR INEQUALITIES

1) The locus of a linear equation of the form ax+by=c is the set of all points whose coordinates (x,y) satisfy the equation. The linear equation can also be an inequality.

2) The set of all points on one side of a straight line, excluding the points on the line itself, is called an open half plane.

3) An open half plane together with the points lying on the bordering line is called a closed half plane.

4) The intersection of closed half planes is called a polygonal convex set.

5) A system of inequalities is a set of two or more simultaneous linear inequalities.

6) Theorem: The points which are simultaneous solutions of a system of inequalities of $\leq$, $\geq$ type, form a polygonal convex set.

9.2 LINEAR PROGRAMMING

1) A finite polygonal convex set is called a convex polygon.

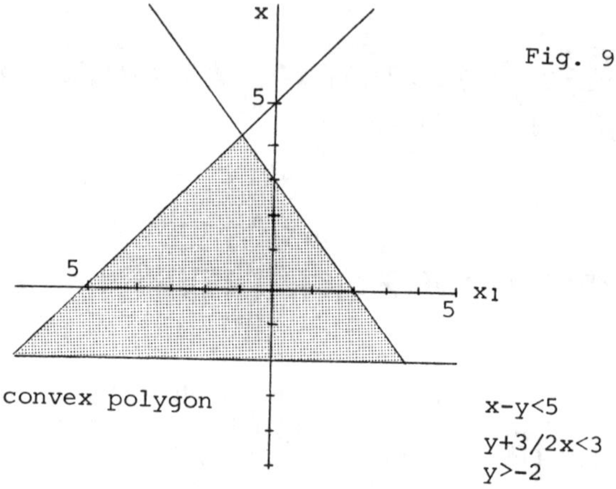

Fig. 9.1

2) For a convex polygon, a boundary point refers to a point lying on a side of the polygon; a corner point refers to a point located at the intersection of two sides of the polygon.

3) Theorem: A linear function ax+by or ax+by+c, and a corresponding convex polygon, takes on its maximum and minimum values at corner points of the convex polygon.

4) Method for finding the maximum or minimum of a linear function over a convex polygon:

   Step 1. Find the corner points of the polygon.

   Step 2. Substitute the coordinates of each of these points into the function.

   Step 3. The largest value resulting from this substitution is the maximum of the function; the smallest resulting value is the minimum of the function.

5) Problems concerning the determination of the maximum and/or minimum of a function are called linear programming problems.

9.3 GAME THEORY

1) A game is a situation in which two or more players take actions which influence, but do not completely determine, the outcome of a single event. The status of the outcome of this event determines whether a certain player wins, loses, or possibly ties with the other player(s).

2) A pure strategy is a complete set of rules which shows a player how to make game decisions. A player may have several different strategies for a particular game.

3) A matrix game consists of two players, each of whom performs certain actions, as specified by the game rules. At the conclusion of the game the loser pays the winner a certain amount of "reward" (money, points, etc.).

4) Let player A have strategies numbered 1,2,...,M and player B have strategies numbered 1,2,...,N for a certain matrix game. Let X_{ij} be the amount the winning player receives when that player chooses strategy i and the losing player chooses strategy j. The game matrix is then the matrix which has M rows (corresponding to the number of strategies of A), and N columns (corresponding to the number of strategies of B), and also has elements X_{ij}:

	B	
A	x_{11}	x_{12}
	x_{21}	x_{22}

5) Let G =

$$G = \begin{bmatrix} g_{11} & g_{12} & \cdots & g_{1M} \\ g_{21} & g_{22} & \cdots & g_{2M} \\ \vdots & \vdots & & \vdots \\ g_{N1} & g_{N2} & \cdots & g_{NN} \end{bmatrix}$$

be a game matrix. A is the player corresponding to the rows of the matrix, and B is the player corresponding to the columns of the matrix. Let $V = \{v_1, v_2, \ldots, v_N\}$ and $V' = \{v_1', v_2', \ldots, v_M'\}$ be vectors consisting of non-negative integers, where $v_1+v_2+\ldots+v_N = 1$ and $v_1'+v_2'+\ldots+v_M' = 1$. Then V and V' are called the mixed strategy vectors for A and B, respectively. Each component in a mixed strategy vector

represents the ratio: $\dfrac{\text{number of times S is used}}{\text{number of repetitions of the game}}$
where S is the pure strategy represented by the row (or column) in the game matrix corresponding to the mixed strategy vector component.

6) Let G be a game matrix with $W = \{w_1, w_2, \ldots, w_M\}$ and $X = \{x_1, x_2, \ldots, x_N\}$ being its associated mixed strategy vectors. If $w_1 = w_2 = \ldots = w_M = x_1 = x_2 = \ldots = x_N = v$, then v is called the value of the game. P and Q are called the optimal strategies for the players if and only if:

   a) $PG \geq W$

   b) $GQ \leq X$

   The value of a game refers to that row and column of the game matrix which represents the pure strategy which will assure each player of at least breaking even.

7) Theorem: If a matrix game has a value and optimal strategies, then that value is unique.

8) Theorem: If G is a matrix game with value v and optimal strategies P and Q, then $v = PGQ$.

9) Theorem: Let G be a matrix game with player A represented by the rows and player B represented by the columns of the matrix. If v is the value of the game, then v is the largest amount which A can be assured of winning, and v is the smallest amount which B can be assured of winning.

10) A matrix game is strictly determined if there is an entry g_{ij} in G that is the lowest entry in the ith row and the greatest entry in the jth column. The value of the game is $v = g_{ij}$.

11) Theorem: If P_1 and P_2 are optimal strategies for player A in a game matrix G, then the strategy

$$P_3 = aP_1 + (1-a)P_2$$

   where $0 \leq a \leq 1$, is also an optimal strategy for A.

12) Theorem: If k is a non-negative number, and G is a matrix game with value v, then the game kG is a matrix game with value kv, and every strategy optimal in G is also optimal in kG.

13) Theorem: Let G be an $M \times N$ matrix game with the value v, and let E be an $M \times N$ matrix, all of whose entries are

1. If k is a constant, then the game $G+kE$ has value $v+k$, and every strategy optimal in game G is also optimal in the game $G+kE$.

14) Fundamental Theorem: If G is any $M \times N$ matrix game, then there exists a value v for G and optimal strategies for each player.

15) If A is an $M \times N$ matrix game, then row i majorizes row h if every entry in row i is at least as large as the corresponding entry in row h. Column j minorizes column k if every entry in column j is at least as small as the corresponding entry in column j.

16) A majorized row or minorized column can be omitted from the matrix game without affecting its outcome.

APPENDIX

FUNCTIONS

BASIC CONCEPTS

Definition of Function
Every relation f such that the conditions $x\ f\ y_1$ and $x\ f\ y_2$ imply $y_1 = y_2$ is called a function (or a mapping).

Instead of writing $x\ f\ y$ and $f \in X \times Y$ we write $f(x) = y$ and $f : X \to Y$

Every element $x \in X$ has a uniquely determined $y \in Y$ such that $f(x) = y$.

If $y = f(x)$ we say that y is the image of x under f and x is a preimage of y under f.

The set Y, in $f : X \to Y$, is called the range of the function.

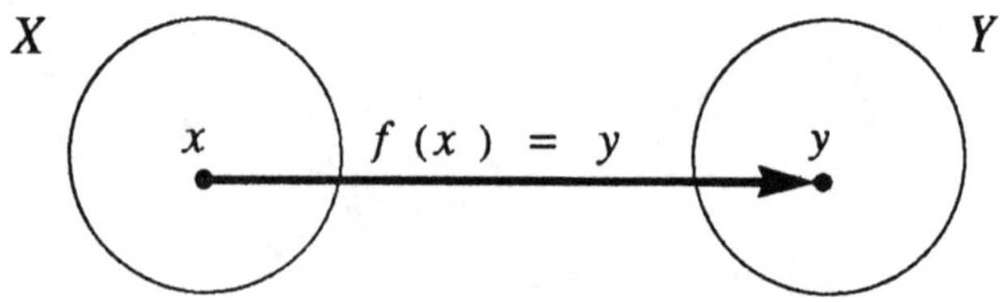

y is the image of x

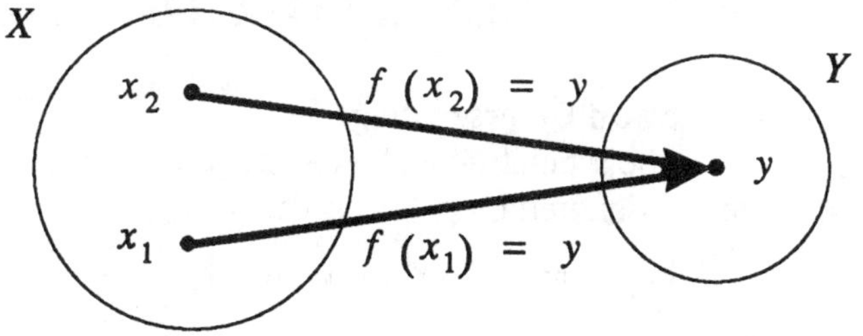

x_1 and x_2 are preimages of y.

Theorem
Let $f : X \to Y$ and $g : X \to Y$ be functions. Then $f = g$ if and only if

$$\forall x \in X, f(x) = g(x)$$

Example:
Let X be a set. The identity function 1_x on X is defined by

$$1_x : X \to X$$
$$1_x(x) = X$$

for all $x \in X$.

Example
Let X and Y be two nonempty sets and let y_0 be a fixed element of Y. Function $f : X \to Y$ such that for all $x \in X$, $f(x) = y_0$ is called a constant function.

Theorem
Let $f : A \to C$ and $g : B \to D$ be two functions such that $f(x) = g(x)$ for all $x \in A \cap B$

Then the union h of f and g defines the function

$$h = f \cup g : A \cup B \to C \cup D$$

where

$$h(x) = \begin{cases} f(x) & \text{for } x \in A \\ g(x) & \text{for } x \in B \end{cases}$$

IMAGES AND INVERSE IMAGES OF SETS

Definition of Image and Inverse Image

Let $f : X \to Y$ be a function and let A be a subset of X. The image of A under f, denoted by $f(A)$ is the set

$$f(A) = \{f(x) : x \in A\}$$

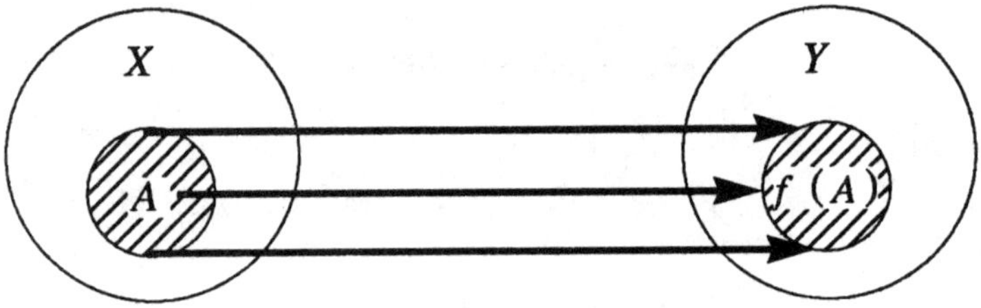

Let $B \subset Y$. The inverse image of B under f, denoted by $f^{-1}(B)$ is the set

$$f^{-1}(B) = \{x \in X : f(x) \in B\}$$

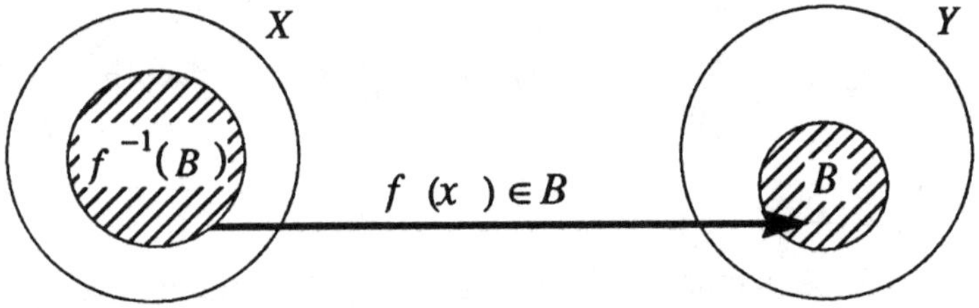

Theorem

Let $f : X \to Y$ be a function then

1. $f(\varnothing) = \varnothing$
2. $f(\{x\}) = \{f(x)\}, \forall x \in X$
3. if $A \subset B \subset X$, *then* $f(A) \subset f(B)$
4. if $C \subset D \subset Y$, *then* $f^{-1}(C) \subset f^{-1}(D)$

Theorem

Let $f : X \to Y$ be a function and let $\{A_i : i \in I\}$ be a family of

subsets of X.
Then

$$1.\ f\left(\bigcup_{i\in I} A_i\right) = \bigcup_{i\in I} f(A_i)$$
$$2.\ f\left(\bigcap_{i\in I} A_i\right) \subset \bigcap_{i\in I} f(A_i)$$

In general, $f\left(\bigcap_{i\in I} A_i\right) \neq \bigcap_{i\in I} f(A_i)$

Example
Let $X = \{a, b\}$ and $Y = \{c\}$, $I = \{1, 2\}$, $A_1 = \{a\}$, $A_2 = \{b\}$.

We define $f : X \to Y$

by $f(a) = f(b) = C$

then $f(A_1 \cap A_2) = f(\varnothing) = \varnothing$

and $f(A_1) \cap f(A_2) = \{C\}$

hence $f\left(\bigcap_{i\in I} A_i\right) \neq \bigcap_{i\in I} f(A_i)$

Theorem
Let $f : X \to Y$ be a function and let $\{B_k : k \in K\}$ be a family of subjects of Y.

Then

$$1.\ f^{-1}\left(\bigcup_{k\in K} B_k\right) = \bigcup_{k\in K} f^{-1}(B_k)$$
$$2.\ f^{-1}\left(\bigcap_{k\in K} B_k\right) = \bigcap_{k\in K} f^{-1}(B_k)$$

Theorem
Let $f : X \to Y$ be a function and let A and B be any subsets of Y.

Then

$$f^{-1}(A-B)=f^{-1}(A)-f^{-1}(B)$$

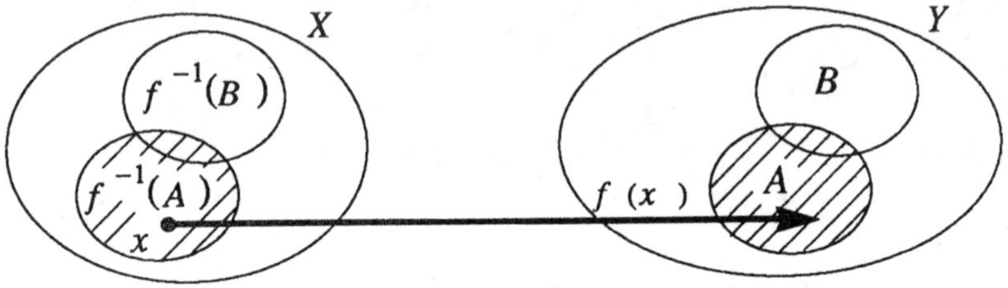

In fact,

$$x\in f^{-1}(A-B)\Leftrightarrow f(x)\in(A-B)\Leftrightarrow$$
$$f(x)\in A\wedge f(x)\notin B\Leftrightarrow x\in f^{-1}(A)\wedge x\notin f^{-1}(B)$$
$$\Leftrightarrow x\in\left[f^{-1}(A)-f^{-1}(B)\right]$$

SURJECTIVE, INJECTIVE, AND BIJECTIVE FUNCTIONS

Definition of Surjective Functions
A function $f: X\rightarrow Y$ is called a surjection (or is said to be onto) if for every $y\in Y$ there exists at least one $x\in X$ such that $f(x)=y$

Function $f: X\rightarrow Y$ is a surjection if and only if $f(X)=Y$

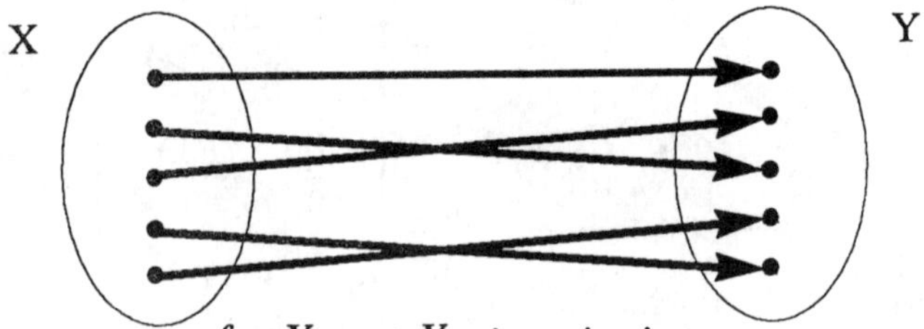

$f: X\rightarrow Y$ is surjective

Definition of Injective Function

A function $f : X \to Y$ is called an injection (or is said to be one-to-one) if for any $x_1, x_2 \in X$

$$f(x_1) = f(x_2) \Rightarrow x_1 = x_2$$

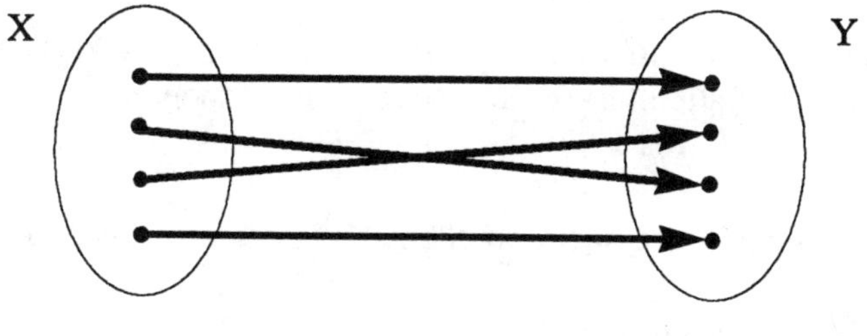

$f : X \to Y$ is injective (but not surjective).

Example

Consider function $f : \mathcal{R} \to \mathcal{R}$ defined by $f(x) = 3x + 2$ this function is surjective. For every $y \in \mathcal{R}$ there exists $x \in \mathcal{R}$ such that $f(x) = y$ This function is also injective.
Function $f : \mathcal{R} \to \mathcal{R}$ $f(x) = x^2$ is not injective because $f(-1) = f(1)$.

Definition of Bijective Function

A function $f : X \to Y$ is called a bijection (or is said to be bijective) if it is both surjective and injective.

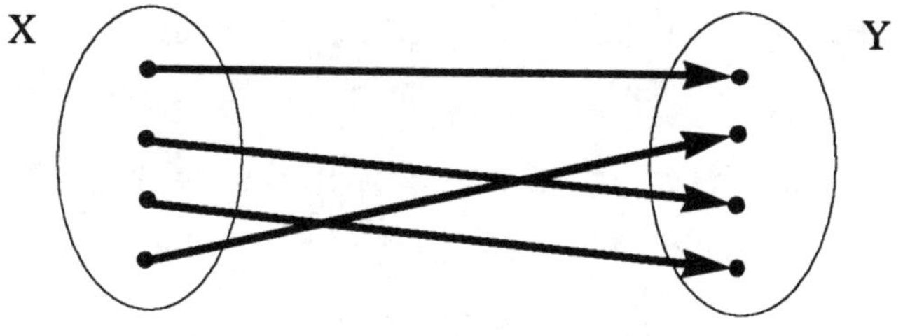

$f : X \to Y$ is a bijection.

Theorem

Let $f : X \to Y$ be an injection and let $\{A_k : k \in K\}$ be a family of subsets of X.

Then

$$f\left(\bigcap_{k \in K} A_k\right) = \bigcap_{k \in K} f(A_k)$$

Suppose R is a relation from X to Y, then its inverse

$$R^{-1} = \{(y, x) : (x, y) \in R\}$$

is a relation from Y to X. Since a function $f : X \to Y$ is a relation, f^{-1} is also a relation. In some cases f^{-1} is a function.

Theorem
If $f : X \to Y$ be a bijection, then $f^{-1} : Y \to X$ is a bijection.

Definition of Extension
Let $f : X \to Y$ and $f_1 : X_1 \to Y$ where $X \subset X_1$.
If for each $x \in X$

$$f(x) = f_1(x)$$

then f_1 is called an extension of f to X_1 and f is called a restriction of f_1 on X. We write

$$f_1 | X = f$$

COMPOSITION OF FUNCTIONS

Definition of Composition
Let $f : X \to Y$ and $g : Y \to Z$ be two functions. The composition of f and g is the function $g o f : X \to Z$ defined by $(g o f)(x) = g(f(x))$ for all $x \in X$.

We have

$$g o f = \{(x, z) \in X \times Z : \exists y \in Y \text{ such that } (x, y) \in f \wedge (y, z) \in g\}$$

Example
Let $f : \mathcal{R} \to \mathcal{R}$ and $g : \mathcal{R} \to \mathcal{R}$ be two functions defined by

$$f(x) = x + 2, \quad g(x) = x^2$$

for all $x \in \mathcal{R}$.

Then $(g \circ f)(x) = g(f(x)) = g(x+2) = (x+2)^2$

and $(f \circ g)(x) = f(g(x)) = f(x^2) = x^2 + 2.$

Hence in general $g \circ f \neq f \circ g$

i.e. composition is not commutative.

Theorem
Functional composition is associative.
That is, if $f: X \to Y$, $g: Y \to Z$, $h: Z \to U$

then $(h \circ g) \circ f = h \circ (g \circ f)$

Theorem
Let $f: X \to Y$ be a function. Then

1. If there exists a function $g: Y \to X$ such that $g \circ f = 1_x$ (where $1_x : X \to X$ is the identity function), then $f: X \to Y$ is an injection.

2. If there exists a function $h: Y \to X$ such that $f \circ h = 1_y$, then $f: X \to Y$ is a surjection.

Inductive Logic

Argument by Analogy

An argument by analogy is an argument based on a comparison or analogy as follows:

(1) A comparison is made between two or more objects with respect to their sharing various properties in common.

(2) All but one of these objects is claimed to have an additional property.

(3) The conclusion is drawn that the other object has the property.

Call the object mentioned in the conclusion the "target entity" and the property mentioned in the conclusion the "target property." Call the other objects mentioned in the premises the "base entities" and the properties mentioned in the premises, except the target property, the "base properties."

The form of a simple argument by analogy can be represented as follows:

Aa & Ba & Ca & ... & Ma	Here, the base properties are A, B, C, ..., M;
Ab & Bb & Cb & ... & Mb	the base entity is a; the target property is N; and
Na / Nb	the target entity is b.

A more complex pattern of argument by analogy simply increases the number of base entities.

The *positive analogy* is the properties the base entities and the target entity share in common: A—M in the form above. The *negative analogy* is the properties not shared between the target entity and the base entities. The *degree of analogy* between entities is greater with larger positive and smaller negative analogy, and smaller with smaller positive and larger negative analogy. The *range* of the analogy is the set of entities involved in the argument by analogy. The *relevance* of one or more properties to another property is the regular or lawful connection between them. For example, horsepower of a car engine is relevant to the car's gas mileage, whereas color of the body is not.

Several rules are common for evaluating the strength of an argument by analogy.

<u>Rule of Relevance</u>.	The more/less relevant the positive analogy is to the conclusion, the stronger/weaker the argument; and/or the more/less relevant the negative analogy is to the conclusion, the weaker/stronger the argument.
<u>Rule of Degree</u>	The greater the degree of analogy, the stronger the argument (other considerations being equal).
<u>Rule of Range</u>	The greater the range of analogy, and the more diverse its members, the stronger the argument.
<u>Rule of Strength of the Conclusion</u>	The stronger/weaker the conclusion, the weaker/stronger the argument. the more/less specific the conclusion, the weaker/stronger the argument.

Induction by Simple Enumeration

Induction by simple enumeration is closely related to argument by analogy. This form of inference underlies such common examples as the following: All emeralds are green, since all emeralds observed so far have been green.

The form of an argument by induction by simple enumeration is:

Instance 1	Aa & Ba
Instance 2	Ab & Bb
...	...
Instance n	Am & Bm
	All As are Bs

Mill's Methods

Mill's Methods are methods for determining the probable presence of causally necessary or sufficient conditions of a given property from observation of instances of events.

The *conditioned property* is the property whose necessary or sufficient conditions are being sought. *Possible conditioning properties* (PCPs) are those properties under consideration as being necessary or sufficient for a given conditioned property. Mill's Methods are ways of selecting necessary or sufficient causal conditions from among the PCPs on the basis of observation of various instances. The methods involve practical applications of the definitions of necessary condition and of sufficient condition.

A property P is a *necessary condition* for a property Q iff whenever Q is present, P is present; Q cannot occur without P's having occurred: if no P, then no Q.

A property P is a *sufficient condition* for a property Q iff whenever P is present, Q is present; Q can't fail to occur when P occurs: if P, then Q.

In applying Mill's Methods, it is assumed that the PCPs include the causally necessary or sufficient conditions for the conditioned property under investigation. It must also be recognized that the conclusions drawn may be overturned by examination of additional instances; these inferences are inductive.

Assume for purposes of illustration a set {A, B, C, D} of PCPs and a conditioned property E.

The Direct Method of Agreement

The Direct Method of Agreement is used to identify a necessary condition for a conditioned property E from among the PCPs. Any property that is absent when the conditioned property E is present cannot be a necessary condition of E. If, where E is present in all observed instances, only C is present in all observed instances, then the other PCPs cannot be necessary conditions for E. Hence, if a necessary condition for E is among the PCPs under consideration, then it is C.

Elimination Principle: A property that is absent when E is present can't be a necessary condition for E.

Factor Tabulation				
Instance	Possible Conditioning Properties			Conditioned Property
	A	B	C	E
1	Present	Present	Present	Present
2	Present	Absent	Present	Present
3	Absent	Present	Present	Present
If one of the PCPs is a necessary condition for E, C is that necessary condition				

The Inverse Method of Agreement

The Inverse Method of Agreement is used to identify a sufficient condition for a conditioned property E from among the PCPs. Any property that is present when the conditioned property E is absent cannot be a sufficient condition of E. If, where E is absent in all observed instances, only C is absent in all observed instances, then the other PCPs cannot be sufficient conditions for E. Hence, if a sufficient condition for E is among the PCPs under consideration, then C is that sufficient condition.

Elimination Principle: A property that is present when E is absent can't be a sufficient condition for E.

Factor Tabulation				
Instance	Possible Conditioning Properties			Conditioned Property
	A	B	C	E
1	Present	Absent	Absent	Absent
2	Absent	Present	Absent	Absent
3	Present	Absent	Absent	Absent
If one of the PCPs is a sufficient condition for E, C is that sufficient condition				

The Method of Difference

The Method of Difference is used to find a sufficient condition for a conditioned property in a particular instance where it is present. That is, the Method of Difference is used to determine from among the conditions present when a particular instance of E occurs, which property is a sufficient condition for E on this occurrence. For example, being blown apart by an explosive is sufficient for death, but when an intact corpse is discovered, that sufficient condition for death will not be the one sufficient for death in this instance, for the person was not blown up; something else was sufficient for death in this case. When a given conditioned property is present, some of its sufficient conditions may be absent. Any sufficient condition present in the instance of interest is sufficient for E in that instance. The Method of Difference is used to determine which of the PCPs actually present on a particular occurrence of E is a sufficient condition for E.

Principle of Elimination: A property that is present when E is absent cannot be a sufficient condition for E.

Let "Occurrence*" be the particular occurrence of E for which a sufficient condition is sought.

Factor Tabulation					
Instance	Possible Conditioning Properties				Conditioned Property
	A	B	C	D	E
Occurrence*	Present	Absent	Present	Present	Present
1	Absent	Absent	Absent	Present	Absent
2	Present	Absent	Absent	Absent	Absent
If one of the PCPs is a sufficient condition for E in Occurrence*, then C is that sufficient condition. Note that B may be a sufficient condition for E, but B was not present on Occurrence*, and so could not have been a sufficient condition for E on that occurrence.					

Combined Methods

The direct method of agreement (to find necessary conditions) may be combined with either the Inverse Method of Agreement or the Method of Difference (to find sufficient conditions) to find conditions both necessary and sufficient.

The *Double Method of Agreement* combines the Direct Method of Agreement with the Inverse Method of Agreement to determine a property both necessary and sufficient for E. First eliminate PCPs that are absent when E is present as necessary conditions. So if any of the PCPs are necessary conditions, they are the remaining ones. Then eliminate the PCPs that are present when E is absent as sufficient conditions. So if any of the PCPs are sufficient conditions, they are the ones remaining. But since those remaining have also passed as necessary conditions, they are both necessary and sufficient conditions for E.

Factor Tabulation					
Instance	Possible Conditioning Properties				Conditioned Property
	A	B	C	D	E
1	Absent	Present	Present	Present	Present
2	Present	Absent	Present	Absent	Present
3	Present	Absent	Absent	Absent	Absent
4	Absent	Present	Absent	Present	Absent
If one of the PCPs is a necessary condition for E and if one of the PCPs is a sufficient condition for E, then C is that property which is both necessary and sufficient.					

The *Joint Method of Agreement and Difference* combines the Direct Method of Agreement with the Method of Difference to determine a property both necessary and sufficient for E.

First apply the Method of Difference to a particular occurrence of E: (a) any property not present at Occurrence* cannot be a sufficient condition for that occurrence of E; (b) any property present when E is absent cannot be a sufficient condition for E for any occurrence of E. The properties remaining are sufficient conditions for E if any properties among the PCPs are.

Then apply the Direct Method of Agreement to eliminate as a necessary condition any property that is absent when E is present to obtain any necessary conditions for E.

Finally, combining the two results, any property that remains is both necessary and sufficient for E if any property among the PCPs is sufficient for E and if any property among the PCPs is necessary for E.

Factor Tabulation					
Instance	Possible Conditioning Properties				Conditioned Property
	A	B	C	D	E
Occurrence*	Present	Absent	Present	Absent	Present
1	Present	Absent	Absent	Absent	Absent
2	Absent	Present	Present	Present	Present
If one of the PCPs is present in Occurrence* is a sufficient condition for E, and if one of the PCPs is a necessary condition for E, then C is both necessary and sufficient for E. Note that B and D might be both necessary and sufficient for E, but they were eliminated because they were not present on Occurrence*.					

The Method of Residues

The Method of Residues separates from a group of causal conditions (sufficient or necessary) those that are already known, leaving the remaining condition (the residue) as the causal condition of E. For example, suppose that the properties ABC are causally connected with the complex of properties $E_1E_2E_3$, and it is known that A is a causal condition of E_1 and that B is a causal condition of E_2. It is inferred that C is a causal condition of E_3 by the Method of Residues. Like the earlier methods, the Method of Residues is a method that involves elimination of all but one factor.

The Method of Concomitant Variation

The Method of Concomitant Variation serves to identify a causal condition (necessary or sufficient) between a conditioned property E and some PCPs by matching variations in one of the PCPs with variations in E. The variations may be direct or may be inverse. In such a case, the conclusion is drawn that if the PCPs include the causal condition, it is the one whose variation changes regularly with changes in E. In the table, "+" and "-" are used to indicate that a property is present to a greater or lesser degree.

Factor Tabulation				
Instance	Possible Conditioning Properties			Conditioned Property
	A	B	C	E
1	Present+	Present-	Present	Present
2	Present-	Present	Present+	Present+
3	Present	Present+	Present-	Present-
If a casual condition is present among the PCPs, it is C, which varies directly with E.				

Factor Tabulation				
Instance	Possible Conditioning Properties			Conditioned Property
	A	B	C	E
1	Present+	Present-	Present	Present
2	Present-	Present	Present-	Present+
3	Present	Present+	Present+	Present-
If a casual condition is present among the PCPs, it is C, which varies inversely with E.				

Probability

Theories of Probability

Truth tables provide a method of determining the truth value of a compound statement given the **TV**s of its atomic components, but gave us no method for determining the **TV**s of the atomic constituents themselves. In the same way, the probability calculus provides a way to determine the probability of a compound statement given the probabilities of its constituents, but offers no way to determine the probabilities of those simple constituents themselves.

There are at least three different theories of probability, each of which provides a way to determine the probability of simple statements. Each theory applies in somewhat different situations.

The *classical* theory of probability (or *logical* probability) determines the probability of a kind of event independently of any observations. The probability of an event E is determined by the formula

$\Pr(E) = f/n$,

where "f" is the number of favorable outcomes and "n" is the total number of possible outcomes. This theory assumes that all possible outcomes are equally probable.

The *relative frequency* theory assigns probabilities to simple statements on the basis of the frequencies with which certain kinds of events are observed to occur. The probability of an event E on the relative frequency theory is given by the formula

$\Pr(E) = f_o / n_o$,

where "f_o" is the number of *observed* favorable outcomes and "n_o" is the total number of *observed* outcomes.

The *subjectivist* theory interprets probability in terms of the beliefs of individual people. The vagueness that attaches to this concept is given a quantitative interpretation via the odds a person is willing to take on a bet.

The Probability Calculus

The probability calculus codifies in a collection of rules and definitions what a quantity must satisfy in order to be considered a probability value. The probability calculus formulates the common core of each of the concepts of probability.

Probability values range from 0 to 1.

If a statement □ is logically true, then Pr(□) = 1.

If a statement □ is logically false, then Pr(□) = 0.

If a statement □ is contingent, then 0 < Pr(□) < 1.

If statements □, Δ are logically equivalent, then Pr(□) = Pr(Δ).

Statements □, Δ are *mutually exclusive* iff {□, Δ} is inconsistent.

Negation Rule: Pr(~□) = 1 Pr(□).

Conditional probability is the probability that one event, Δ, will occur given that another event, □, has occurred, Pr(Δ given □):

Conditional probability: Pr(Δ given □) = Pr(Δ & □) / Pr(□), where Pr(□) > 0.

Independence: □ and Δ are independent iff Pr(Δ given □) = Pr(Δ).

The *special conjunction rule* is used to calculate the probability that two events will occur when the events are independent of each other:

Special conjunction rule: If □ and Δ are independent, then Pr (□ & Δ) = Pr(□) × Pr(Δ).

The *general conjunction (disjunction) rule* is used to calculate the probability that either one or the other of two events will occur whether or not they are independent of each other.

General conjunction rule: Pr(□ & Δ) = Pr(□) × Pr(Δ given □).

The *special disjunction rule* is used to calculate the probability that either of two mutually exclusive events will occur:

Special disjunction rule: If □ and Δ are mutually exclusive, then Pr(□ VΔ) = Pr(□) + Pr(Δ).

The *general disjunction rule* is used to calculate the probability that either one or the other of two events will occur whether or not they are mutually exclusive:

General disjunction rule: Pr(□ VΔ) = Pr(□) + Pr(Δ) Pr(□ & Δ).

Bayes' Theorem may be used to calculate the conditional probability of □ given Δ from the probability of Δ given □. For example, the probability that Lassie is a dog given that she is a collie is 1, but the probability that she's a collie given that she is a dog is less than 1.

The special version of Bayes' Theorem applies where events are mutually exclusive and jointly exhaustive, as Δ and ~Δ are.

Bayes' Theorem:

$$Pr(\Delta given \square) = \frac{\{Pr(\Delta) \times Pr(\square\, given \Delta)}{[Pr(\Delta) \times Pr(\square\, given\ \Delta)] + [Pr\ (\Delta) \times Pr(\square\, given{\sim}\Delta)]}$$